Andrew McDeere

BACKYARD CHICKENS
and
RAISING GOATS

A Complete Guide to Raising Chickens and Goats

Copyright © 2019 publishing.

All rights reserved.

Andrew McDeere

No part of this publication may be reproduced, distributed or transmitted in any form or by any means, including photocopying recording or other electronic or mechanical methods or by any information storage and retrieval system without the prior written permission of the publisher, except in the case of brief quotation embodies in critical reviews and certain other non-commercial uses permitted by copyright law.

Table of Contents

Definition of 'poultry.' 10

BEGINNER'S GUIDE TO RAISING BACKYARD 14

    Starting Out Q&A 15

    Planning and Buying Your Chickens 17

    Financial Statements 17

    How Many Chickens Should I Get? 18

    Where Do I Get My Chicks? 19

    What Should I Look Out For? 19

    Chicken Coops 101 20

    Basic Shelter Requirements 20

    Adequate Amount of Space 20

    Temperature Control 20

    Nesting Boxes 21

    Roosts 21

    Outside Roaming/Pen 21

    Security 21

HOW TO RAISE CHICKS 23

    Brooding Box 23

    Bedding 24

    Warmth 24

    Food 25

    Water 26

    Hygiene 27

    Security 28

HOW TO RAISE CHICKENS 30
Water 30
Food 31
Hen Morning and Evening Routines 31
Common Chicken Problems 32
Molting 32
Stopped Egg Laying 32

Reasons Why Your Chickens Stopped Laying Eggs 33
Broodiness 40
What Exactly Is A Broody Hen And How To Stop It? 40
What Is A Broody Hen? 41
How to Stop a Broody Hen? 43

BEST BEGINNER CHICKEN BREEDS 47
Why Do You Want Chickens? 47
Typical Chicken Characteristics Beginners Look For 51

BEST EGG LAYING CHICKEN BREEDS 52

SURPRISING RULES FOR FEEDING CHICKENS 58
What Should You Feed Chickens? 58
How to Feed chickens 60
How much are you supposed to feed them? 60
Water for Your Hens 62
Feeding Chickens Table Scraps 62
5 Healthy Treats 64
What you Shouldn't Feed Them 64
What Can Happen if Their Diet Isn't Right? 65

TOP 13 BEST EGG LAYING CHICKEN BREEDS 67

WHAT TIME DO HENS USUALLY LAY? 82

HOW COLD IS TOO COLD FOR BACKYARD CHICKENS 85
- Signs of Cold Weather Stress 86
- A Well Designed Coop 86
- What About Sick or Injured Chickens and the Cold Weather? 90
- Keeping the Drinking Water from Freezing 91
- How Cold is Too Cold for Backyard Chickens? 92

THE MOST RARE CHICKEN BREEDS 93
- Dong Tao 93
- Ayam Cemani 95
- History 97
- GENERAL 99
- GROOMING 99
- DIET OR NUTRITION 100
- SOCIALIZING THE BIRD 100
- Onagadori 102
- Polverara 107
- Ixworth 108
- Naked Neck 109
- Golden Campine 113
- Where To Find These Birds To Add To Your Flock 115
- CARING FOR THE BIRD(s) 115
- GENERAL 115
- GROOMING 116
- DIET AND NUTRITION 116
- SOCIALIZING THE BIRD(s) 117
- Vorwerk 119

- Modern Game 120
  - HISTORY 122
- **TAKING CARE OF YOUR CHICKEN FLOCK** **125**
  - Vaccinations 125
  - De-worming 127
  - Types of worms: 130
  - Tapeworm 130
  - Gapeworm 131
  - Round-Worm 131
  - TAKING CARE OF YOUR FLOCK 132
  - Mites: 139
  - Lice: 140
  - Fleas 141
  - Ticks 142
  - Bathing your chickens 145
  - Clipping nails 147
  - Clipping wings 148
  - Dust baths 149
  - The hot summer months 152
  - The colder winter months 162
  - Foraging 167
  - Once a month 172
  - Each six months 172
  - Once a year 172
- **CONCLUSION** **173**
- **Raising Goats** **177**
- **Your Guide To Goat Farming** **185**

- Raising Goats As Pets 188
- Various Types of Goats 192
- Multiple Uses of Goat 198
- Guide to Raising Goats 203
- Why Should You Raise Goats? 212
- How to Raise Goats 215
- Caring For Goats 233
- Goat Meat Nutrition 240
- A Quick Guide to Goat Health 243
- Goat Diseases 251

Andrew McDeere – Rosanne Fox

BACKYARD CHICKENS

For beginners

"The Essential Backyard Chickens Guide for Beginners: Choosing the Right Breed, Raising Chickens, Feeding, Care, and Troubleshooting."

Copyright © 2019 publishing.

All rights reserved.

Andrew McDeere

No part of this publication may be reproduced, distributed or transmitted in any form or by any means, including photocopying recording or other electronic or mechanical methods or by any information storage and retrieval system without the prior written permission of the publisher, except in the case of brief quotation embodies in critical reviews and certain other non-commercial uses permitted by copyright law.

INTRODUCTION

Raising backyard poultry can be rewarding not only for the production of consumable eggs and meat but also for companionship and personal pleasure. Providing veterinary care to these clients and their birds can be equally rewarding as long as the veterinarian has a strong understanding of federal, state, and local laws and regulations that affect backyard poultry and has a working knowledge of common diseases and treatments for these species.

Definition of 'poultry.'

The word 'poultry' should be treated as referring to any avian species that has the capacity for its meat, eggs, or other sections (e.g., offal, feathers, manure) to join, directly or indirectly in the human food chain, regardless of the actual use of individual birds of each species. The Code of Federal Regulations (CFR) describes 'major' animal species like dogs, cats, horses, goats, pigs, chickens, and turkeys. All others are known to be "minor" animal species, such as ducks, geese, game birds, pigeons, etc. Demographics It is widely recognized that backyard poultry is growing in popularity. This is not shocking, considering the national trend towards locally produced food and organic food. Poultry breeding can also be socially enriching and instructional. According to the U.S. Department of Agriculture (USDA) National Animal Health Monitoring System (NAHMS) Poultry 2010 study "Urban Chicken Ownership in Four U.S. Cities," which surveyed urban chicken ownership in Chicago, Los Angeles, Houston, and New York City, 0.8% of all households in these cities owned chickens and 4.3% owned single-family homes on one acre or more. About 4 percent of households that did not have chickens at the time of the study were planning to own chickens in the next five years. With the recent steady decline in parrot possession, there is no question that poultry veterinary patients will become more prominent in the coming years.

Birds The majority of poultry birds fall within a few avian orders, including Galliformes (e.g., chickens, turkeys, quail, pheasants, grouse, and guinea fowl), Anseriformes (ducks, geese, swans) and Columbiformes (pigeons and doves). Ratites such as ostrich and emus should also be regarded as poultry animals. Storey's Illustrated Guide to Poultry Types 2007 is an excellent resource for learning about different types of poultry.

Domestic chicken, Gallus domesticus, is widely believed to be a descendant of red jungle birds from geographical areas such as southern China and Indonesia. At the same time, genetic cross-breeding may have occurred with other avian species, such as gray and Sri Lankan jungle birds. Currently, there are hundreds of chicken types around the world, classified by category (geographical area (e.g., American) and, for bantams, by physical traits such as Single Comb), variation (based on physical characteristics such as color and comb type), and stress. Chickens are often categorized as regular breeds (usually larger and more common and commercially bred for eggs and meat such as bared rocks or Rhode Island Reds), bantam breeds (smaller breeds such as silkies) and heritage breeds. Consumer varieties shall contain broilers and plates. See American Poultry Association Perfection Guidelines (http:/www.amerpoultryassn.com) for more information about chicken breeds.

Regulation and biosecurity Plant laws and regulations. Although the commercial poultry industry is tightly regulated by federal and state legislation, poultry in the backyard is generally excluded. Regulation of backyard poultry flocks is usually limited to local regulations, and some municipalities do not have any regulations at all.

Local regulations may regulate ownership by requiring permits or licenses, or by dictating care and husbandry, such as defining limits on the size of the property and the available coop space per bird, and through the noise, odor and cleanliness nuisance clauses.

Biosafety or quarantine; Note that "an ounce of prevention is worth a pound of cure." Biosafety is the cheapest and most effective way of restricting and stopping the introduction and dissemination of infectious agents to and within a backyard flock. There are three main biosecurity principles: access management (e.g., closed definition of the flock, enclosure, and confinement), animal health management (e.g., good record-keeping, veterinary care) and environmental management (e.g., sanitation or disinfection, pest control).

Backyard poultry inherently poses a risk to the commercial poultry industry for the spread of highly pathogenic diseases such as exotic Newcastle disease (END) and highly pathogenic avian influenza (HPAI). The vet plays a key role in the early detection and notification (reporting) of potentially serious contagious diseases.

BEGINNER'S GUIDE TO RAISING BACKYARD

Chickens Raising chickens can be a lot of stuff: relaxing, satisfying, enjoyable, and for beginners, maybe a little anxious!

There are tons of information on raising chickens and chicks, and it's hard to sort through it all to find out what's true, what's wrong, and what's odd.

In this definitive guide, I've brought everything together

Starting Out Q&A

Before you get your chickens or chicks, you need to ask yourself a few questions: why are you raising them? Eggs, meat, or pleasure?

- Where are you going to get them?
- Are you prepared to spend some time with them?
- Are you able to 'muck out' their coop when it's needed?
- Who's going to take care of them if you go on vacation?
- Are you allowed to have chickens, if so, how many?
- Such questions could be

Choosing the Best Breed of Chicken Here, in the twenty-first century, we have a beautiful selection of chickens to choose from.

How many different races are there?

No-one really knows, but it's estimated to be in the hundreds. There are birds that have been specifically bred for enhanced egg laying, quality meat, fights, and plumage.

While there are several numbers of chicken varieties,

It's wrong! Unfortunately, you're going to be disappointed if you want egg layers and have bought a bunch of Sultan chicks because they look pretty. They're going to lay an egg a week if you're lucky!

Planning and Buying Your Chickens

It's time to take the plunge! You want chickens, and you know which breed you want, but you're not sure where to start.

For the first time, when you get chickens, you have a couple of different choices.

We will look at the good and the bad of each option. You can buy hatching eggs, chicks, poles, or adult birds. Each choice has its merits, but it's really about what you think is best for you.

Financial Statements

Hatching Eggs: These are fertilized eggs that <u>you</u> need to incubate. If you are new to chickens, I don't recommend that you get hatching eggs unless you really know what you are doing. Although incubation is fairly straight forward, there's definitely an art to it.

Chicks: This is the most used and wise choice for novices. You can select which breed(s) you want and when you want them. You typically get chicks at one day old.

Pullets: Pullets are birds aged between four to six months. The chicks have been reared to adulthood, and are usually sold at point of lay, meaning the pullet is about to lay her first egg anytime soon!

Adults: Adult hens are more difficult to come by as breeders like to move birds out before they get too old since they eat more. A common source of adult hens is animal shelters or rescue sanctuaries.

How Many Chickens Should I Get?
You can generally average out how many chicks you will need. If your birds are for eggs only, then you just need to think how many eggs do you use in a week currently?

One hen will average four to five eggs a week. Throw in a couple of extra chicks for 'just in case, and you have your number!

For example, if you want 16 eggs a week, you would need six hens (4 would normally do this many eggs, but I've included 2 'just in case' chicks).

Where Do I Get My Chicks?

The best place for beginners to buy their chickens from is a local farmer, hatchery, or farm supply stores.

Though, if you want to purchase your chickens from further away, the USPS has been shipping chicks for about one hundred years and will ship chicks, which you can purchase online.

What Should I Look Out For?

All birds should have clear, bright eyes. They should be curious about their environment and you. Feathers or fluff should look clean with good coloring. If a bird, regardless of its age, exhibits any of these signs, you should avoid buying it.

- Sleepy, lethargic
- Hunched into a ball
- Sitting by itself
- Reluctant to move
- Any nasal/eye discharge
- Blocked vent

Chicken Coops 101

Chickens are not very strict when it comes to housework.

They don't need running water, power, and carpeting. A modified basic wooden box is in a pinch, but there are few crucial things you need for your flock to ensure their safety and well-being.

Basic Shelter Requirements

This is the most important need of all, a spot where they can get out of the blissful heat, howling wind and blowing snow. The coop has to be water-resistant because there's nothing more unpleasant than a wet hen.

Adequate Amount of Space

Adequate habitat for birds to co-exist comfortably is important. If they're crowded together, they're likely to start anti-social behaviors like picking and pecking each other. Winter is the worst season for these behaviors; hens get restless and make trouble.

Temperature Control

Ideally, the coop should be cool in the summer, warm in the winter. Proper ventilation of your coop is important when it comes to temperature control. A good flow of air will keep the coop at an optimum temperature for your hens. You need to install more ventilation holes if you find it's too warm.

Nesting Boxes

For nesting boxes, you'll need about one box for every three hens, but it never hurts to have more. There's always one particular package they're going to squabble over, so more is good.

Roosts

You'll need about one box for every three hens with nesting boxes, but it never hurts to have more. There's always a certain package they're going to squabble over, so more is good.

Outside Roaming/Pen

In addition to a secure coop, your chicken will also need access to some outside space, whether it is contained or free of charge.

Security

A strong and formidable co-op will keep your girls safe at night. Don't imagine, even for a minute, that wolves won't be brazen enough to creep into your yard and try to kill your birds, they're going to, and they're going to do so, with catastrophic results for your birds and heartbreak for you.

HOW TO RAISE CHICKS

Now that you've done your research, and settled on your type, what's going to happen now?

If you've bought them online, the website you've ordered is likely to carry all the items you need to make your kids back.

If you're buying from a farm store, make sure you've got everything you need for the chicks in advance. Farm stores usually sell a large quantity of chick related goods.

Make sure you know what to do; otherwise, you could be parting with hard-earned cash for something you don't want or need.

It helps to break things down into needs and extras, so that's what I've done here for ease of reading.

Brooding Box

This could be as easy as a cardboard box. It's got to be tall enough to stop the chickens from running down. It's got to be big enough for a food dish, a water dish, and chicks.

There is also a need for cold proof drafts that can kill chicks very quickly.

Brooding boxes come in all shapes, sizes, and costs. If you're not sure if you're going to be brooding your chicks again, get the cheapest brooder to start with, you can upgrade yourself later.

Bedding

Something soft like a pine shaving. They are sold in small bales at most farm stores and are affordable.

If your embroiderer box has a sticky floor (like plastic), place some paper towels under the bedding so that they can stay and hold up well; otherwise, they may have trouble standing or moving.

Warmth

Chicks require a lot of energy.

They don't have true feathers until they're around 6-7 weeks old, so they can't regulate their own temperature, so you need to help them.

You can use a heat lamp or heat plate ' Smart chicken.'

They will need to be warmed for about 6-7 weeks or until the ambient temperature is approximately the same as that of the embroiderer.

The temperature at the chick stage will have to be 95F during the first week. This reduces by 5 degrees per week until the ambient temperature is reached.

How do you know if it's warm enough?

If they're all huddled in a group, they're too cold; if they're scattered to the sides of the embroiderer, they're too hot; if they're clustered all over, they're just correct. A thermometer will help you with this too, but base your judgment on the behavior of the chicks.

Food

Chicken food comes in a wide range of choices that can be overwhelming, so here's the scoop. It is advised that you feed your chicks as follows:

- 0-8 weeks: 18-20% starter feed crumbles
- 8-14 weeks: 16-18% starter/grower
- 15-18 weeks: 16% finisher
- 18 weeks upward: 16% layer feed

Chick feed can come as medicated or un-medicated.

Medicated feed is medicated with a coccidiostat that helps protect them from coccidiosis, a terrible disease.

If your chicks have been vaccinated at the source for coccidia, do not use medicated feed.

Chicks are messy; they're going to scratch their food all over the floor, pop in it, and get their bedding in it, so you need a feeder that's going to remove some of that mess.

Before they start eating greens such as short grass and dandelions, they'll need a little chick-grit dish to support digestion and make sure they don't get a seed.

Water

Water is essential to the well-being of all creatures. Chicks are no exception. Water should be at Goldilock's temperature—not too warm, not too cold, but just right.

You're going to have to dip the beak of each chick in the water at first so that they know where it is; after that, they should all be able to find the water dish. Do the same thing with a food dish, too.

If your chicks are only a few days old, you'll need to add some clean pebbles or marbles to the water dish so they can't fall in and drown. After a week or two, you could cut them because the chicks are going to be strong enough not to sink themselves.

For the first few days, you could apply an electrolyte/vitamin supplement to the water to get them off to a good start.

Change the water frequently (several times a day) as they kick bedding etc. into the water on a regular basis.

Hygiene

Whoever said' cleanness is next to godliness' never kept chickens. I told you that the chicks are messy, so you're going to be the room service for them!

It is very important that their brooder, feeder, and watering area are kept clean. You need to remove the poop daily, change the litter as often as you need it. Once it gets wet, it has to be changed.

Remember, the brooder is very warm, there's poop, and it's wet, it's the perfect breeding ground for bacteria.

Wash and sanitize the feeder and waterer at least every other day. If your chickens are as dirty as mine, you're going to have to throw out a decent amount of food. Once they're pooping in the feeder, the food goes out.

Don't forget, of course, to sanitize and wash your hands before handling food or drinks!

Security

Your brooder full of chicks needs to be safe from wolves, and I also consider house pets as pests.

When you have them in your home, you'll need to make sure that Fido and Fluffy can't get to those little balls of fluff, maybe put them in a separate room or a safe lock on the package.

Seek not to have them in places such as the bathroom, the dining area, and the kitchen. They're stirring up a lot of dirt and dander, people who are allergic to dust may have issues with them at home.

If you're planning to keep the chicks in the outbuilding, you need to be able to exclude any predators; you may have in the area.

Rats, like a chick snack, like foxes, weasels, raccoons, and a host of other carnivorous creatures.

They're going to need you to take intensive care of them until they're around 12 weeks old. Some people say that sooner, but I'm wrong on the cautious side.

Going Outside for the First Time

If you're thinking of putting them out every day for a few hours, you'll need to have something like a dog crate or even a mini-chicken run for them.

Of reality, it's going to have to be animal proof, and that means hawks and owls, as well as digging animals and foxes.

They're going to need a shady area where they can escape from the sun, keep food and water cool.

HOW TO RAISE CHICKENS

So now your chickens have evolved into true chickens!

Managing your adult flock may sound like it's complicated, but it's pretty simple to do. The hen is doing all the job, and you're taking care of her needs. Nonetheless, there are some things you should know before you get confused by hiccups in the process.

Water

Water is essential for all living things, and chickens are no exception.

A hen drinks a cup of water every day. She's going to take frequent small sips all day. Too little water can impact egg production, among many other issues, so make sure they have plenty of water.

There are about 15 cups of water in one US gallon, so if you've got a lot of birds, you'll need a few drinkers for them. As an example, I have about forty birds, and I put out four drinkers in various places to make sure they all have access to water.

You can put the water in any kind of plastic container, but the easiest way is to buy a drinker.

Food

Including water, the other important item chicken wants is food.

Giving the chickens the right food will keep them happy and turn them into an egg-laying unit. If you give them the wrong food, it can lead to all sorts of issues, including abuse and weight loss.

Hen Morning and Evening Routines

Unfortunately, most people live busy lives and don't have to tend their chickens all day.

In the morning, you're going to have to let your chickens out of the coop, check on their food and water, and take a general look around and make sure everyone's ok.

It's time to start the evening routine when the sun comes down. This will include, safely lock the girls inside the coop and also collect eggs (if you haven't already done so).

For example, this is the sweet minimum of caring for your women. There will also be regular activities, such as washing the coop and tending the nesting boxes.

Common Chicken Problems

Unfortunately, at some stage in the life of your chickens, they're likely to have some kind of concern, whether it's broodiness, pests, or abuse.

There are a number of common problems that occur when you have chickens. If you're not ready for them, they may seem intimidating and daunting.

Molting

Molting is the process of losing all old, worn-out feathers and replacing them with new plumage. It's going to happen to all birds, including roosters. Some birds may take up to two years to complete a molt, but a humble chicken is usually done in three months.

Stopped Egg Laying

We're all in love with our feathered friends, but one of the main reasons people hold is because of the chickens.

Reasons Why Your Chickens Stopped Laying Eggs

There are lots of reasons why the chickens might have stopped laying, but you don't need to rush out and buy supermarket eggs right now!

Today, we're going to look at the most common reasons that your chickens have stopped laying and what you can do to get them laid again.

1. Their food: The most common reason why the chickens quit laying is that there's something wrong with their diet. Have you recently changed your diet or even changed the brand of pellets that you feed to your chickens?

We once decided to stop feeding our chickens with layers of pellets and instead feed them with maize. Maize is just ground the corn.

When feeding the girls layers of pellets, we were getting an average of 9 eggs a day, and after feeding them Maize for a matter of days, we were only getting 4-5 eggs a day!

Yikes- This is because maize does not contain a lot of protein, so chickens require about 20 grams of protein every day to keep laying eggs.

Just note, whatever you feed your chickens, they need a well-balanced diet to make sure their bodies are capable of producing eggs.

If you're feeding your girls loads of pellets and they're still unable to lay down, consider giving them protein-rich foods such as pumpkin seeds, oats, and mealworms.

Water is another often overlooked aspect of their diet. When chickens don't have access to fresh water all day, you can say goodbye to your eggs.

2. Not Enough Daylight: So you've made sure your girls get plenty of protein and fresh water, but there's still no eggs in sight. Sometimes, it could just be the wrong time of the year for your hens to lie down.

Your chickens need plenty of natural daylight to lay eggs-at least 14 hours a day and 16 hours is even better.

This means that during the winter, when in the US, natural daylight can drop to less than 9 hours a day, your girls would need another 5 hours of daylight to lay eggs.

The solution is to place an artificial light in their coop and set it to an automated timer. This will certainly keep the egg production up, but it's something we're not going to do.

There's a reason why hens don't lay as much in the winter, and their body needs to rest and recover for the next year. And if you don't give them time to recover their bodies, you'll do more harm than good in the long run.

It's not all bad news though, the chickens aren't supposed to stop laying entirely, and you're supposed to get the odd eggs, but that's it.

3. Broody Hens: So your girls are well fed, getting plenty of sunlight, but they're not laying down yet. It's time to give them up and get a new herd, just a joke!

You might have a brooded hen, and in this case, she won't lay eggs, no matter how much protein or sunshine you give her.

When a hen gets tangled, she needs to hatch her own chickens, so she's going to sit on top of her eggs for 21 days before they hatch. During this 21 day cycle, she won't lay any eggs, not good. There are clear signs to look out for if your hen is bridal: she'll be sitting in the nest box all day.

She's going to become very protective and prevent anything from getting close to her nest.

She's going to remove her breast feathers to heat the eggs out of her body.

When you think your hen is bridal, learn how to avoid my bridal chicken.

4. New additions to the flock: So you don't have a broody hen, but you still don't see any eggs. Have you recently moved your chickens or introduced new chickens to the flock?

Chickens enjoy routine, and the slightest disruption in their schedule usually results in their laid off.

The most frequent daily disturbance they undergo is when they're relocated. This could be when they're delivered to your home after you've bought them, or when you've decided to move their coop.

Chances are that you purchased your chickens as pullets, so they weren't feeding before they came anyway. But if you moved their coop, they're not going to be happy with you!

Give them a couple of days to get around, and they're supposed to start laying again.

If you've just introduced new chickens to the flock, this can also disrupt their routine and egg-laying. When new chickens are introduced, there tends to be some shoving and jostling for the first few days when a new pecking order is established. During this time, they're not going to lay eggs, but again, after a few days, they're going to start laying again.

5. Some Breeds Don't Lay As Many Eggs: Other breeds just don't lay as well as others, and we sometimes forget that, particularly when we hear about how large other people's eggs are.

Breeds such as Rhode Island Reds or Buff Orpingtons may lay more than 200 eggs per year. Whereas other species, such as Ameraucanas and Silkies, are known to lay fewer than 100 eggs per year.

If you're unsure how many eggs your chicken breed should lay every year, this beginner's guide to chicken breeds should help.

6. Old Age: So you have a Red Rhode Island, which is supposed to lay more than 200 eggs a year, and they've just stopped laying.

Unfortunately, as chickens grow older, the amount of eggs they lay slows down. Look at the picture below, and you can see that you normally only get about three years of good chicken egg-laying.

If your Rhode Island Red had laid 200 eggs in their first year, they would have laid around 168 eggs in their second year, 128 eggs in their third year. This number will continue to fall to around 40 eggs in the tenth year.

If your chickens get a little older, then a drop in egg laying is perfectly natural and expected.

There's nothing you can do about it, and it's just nature's way when the chickens mature.

7. Illness: If you have a well-fed, well-established young chicken, it has plenty of natural daylight, and they suddenly stopped laying, there's the possibility that they're ill.

Colds: signs to watch out for are slimy eyes, and they're running around with their beak open, so they can't breathe through their ears. Make sure you isolate any chicken you think might have a cold so that it doesn't spread to the rest of the flock.

Parasites: contain lice, mites, and worms. You'll notice your chickens are going pale, and they're not going to stop itching themselves. The easiest way to treat any parasite is to spray chicken coops and chickens with a chicken cleaner. Something like Johnson's Poultry Housing spray is supposed to do the trick.

Molts: A lot of people misinterpret the signs above as a virus when it's just chicken molding. Chickens mulch each year, and it may take about 6 to 12 weeks for them to grow new feathers, they will not lay eggs during this time period.

If you want to keep track of how many eggs your chicken has laid, this spreadsheet should help. You can either fill it in or print it off on your desktop or stick it up elsewhere.

Broodiness

What's a bridal hen? You're going to know when you see it! She's going to sit in the nest all the way. If anyone comes near her, she's going to grumble, squawk, and puff herself out. Maybe she's going to give you an all-powerful cock too. What's a broody hen, and how to avoid it?

What Exactly Is A Broody Hen And How To Stop It?

Your hen squawks each time you touch her, and she doesn't leave her nesting box, what's up, is she sick? Far from it, and the odds are that she's just bridal and wants chickens. If you're not planning to have chicks, this can be problematic because the hen in question will stop laying eggs.

Whatever is the case, you can be sure that if you want chicks and need a bridal hen, there won't be one in sight, but the day you don't want a bridal hen, it's the day you will see one!

Let's take a look at how to spot a bridal hen and want you to do something to keep her from being bridal.

What Is A Broody Hen?

The broody hen is a hen that needs their eggs to hatch. She's going to sit on top of her eggs (and others she's stolen) all day long trying to hatch them. Obviously, if there's no rooster involved, the eggs won't be viable, and for the rest of her life, she can sit on top of the eggs so that they won't hatch!

There is no exact science as to what makes a hen go broody it's a combination of its hormones, instinct, and maturity.

If you've never seen a bridal hen before you might ask, how do you know if a hen is bridal or not? Believe us, once you've seen the signs, you won't have any delusions of becoming a broody hen.

She's going to stay in her nest all day, and we mean, all-day

She's not even going back to roost with the rest of the chickens at night.

She's going to become very territorial about her nest. This means puffing out her wings or squawking at anything that's trying to get near to her.

She's going to peck and try to bite you if you're trying to move around here, so make sure you wear gloves if you need to move her around.

She may also pick out her breast feathers so that the heat from her body passes through to the eggs.

If you want to raise chicks, then having a bridal hen is perfect-after all, it's nature's best incubator. But, if you don't want chickens, it's troublesome to have a bridal duck. Not only will your bridal hen stop laying, but the worst of all, it can cause other hens to turn bridal as well, say goodbye to your egg production!

You could leave her to 'breast,' and after 21 days (which is when the chicks will hatch if the eggs are fertile), she would take advantage of it, but in our experience, they won't, and they need to be 'split.'

So how are you going to break a hen out of her broodiness?

How to Stop a Broody Hen?

The best way to stop a broody hen is because it never happened in the first place, and there are a few things you can do to reduce the chances of your hens becoming broody.

The first thing to do is to remove the eggs from the nesting box as soon as they have been laid. Second, don't let the hens get into the nesting box after the eggs have been laid that day. Now, unless you're around your girls 24/7, both of these options aren't very practical, and you're probably going to find yourself with a bridal hen at some point, so what are you doing?

Well, you've got a lot of options, and you can break her turmoil without causing any emotional damage to her, so don't worry! Let's look at some of the easier options that should work in most situations.

Cut her from her nest box.

Take the broody hen out of her nest, and leave her off with the rest of the chickens in the field. You can do this at the same time as you feed them to maximum effect. Also, as we've already noted, broody hens can bite, so make sure you wear gloves when you do this. Keep an eye on the hen, because she might be going straight back to the nest box. Make this move many times a day to try to destroy it.

Block the nesting box.

She keeps coming back to the nesting box for a few days. It's time to ramp up the ante. Cut her from the nesting box, as you've already done, except once she's out of the nesting box, she's in just hammer a piece of wood to the door. Take the nesting straw out of the package to further dampen her morale just in case she comes back in!

Make her roost again

If she's still bridal, you have a stubborn girl, but don't worry, we still have some more tricks on our sleeves. Just as it's going dark and your hens are going back to the coop to roost, take your broody hen out of her nest and position it with the other chickens roosting. Chances are she won't be bold enough to risk moving back to the nesting box in the night.

Use frozen vegetables

We've always managed to break our hen's broodiness at this point, but other backyard chicken owners haven't been so lucky, so what else can you do? I've heard a few people put a bag of frozen vegetables under their chicken.

We do this because, when the hen is brooded, their body temperature increases so that lowering it (with frozen vegetables) often sends a message to their brain that they are no longer brooded.

Bring out the Broody Buster.

I'm sure your hen s no longer bridal at this point. If this is the case, there is one option left, a broody cage! Don't think about it being less exciting than it looks. You'll need a wire bottom cage for this. You can use a dog/cat carrier to take the bottom out and cover it with chicken wire. Make sure there's nothing in the cage except food and water- that means no bedding.

Bullying.

The pecking order is called for a reason. Each bird in a flock will have its own location. Those at the top are the first to eat; those at the bottom are the last to eat. It's a very simple but effective hierarchy, so that all members know their position.

Bullying happens to a small degree every day because of this. If a chicken comes out of turn, she'll get a quick peck to the head to remind her of her status.

Predators.

Even if you live in the middle of the city, there's a chicken predator in your neighborhood. Foxes, coyotes, raccoons, and the baby dog down the road are all going to want chicken dinner, and these are just field animals.

The key to your flock's safety is the safety and awareness of predatory animals and the area in which you live.

BEST BEGINNER CHICKEN BREEDS

Choosing the right chicken breed as a beginner can be the difference between thoroughly enjoying the moment with your chickens and wondering why you ever needed chickens in the first place.

Most people don't know exactly what they want or need when they start looking at the breeds, so that's why we decided to write this article today.

We will address exactly what you need to think about before you decide on your breed, explain what characteristics beginners should look for, and then recommend our top 5 picks for the best egg-laying chicken breeds.

Why Do You Want Chickens?
Before we go any further, we'll ask you a few questions, be sure to either write down your answers or keep them in your head.

First of all, why are you trying to keep chickens? Were you looking to keep chickens for eggs, food, pets, or a mixture of these purposes (though the poultry and pet choices may not go too well together!)?

Many chickens are exceptionally good egg-laying (Rhode Island Reds and Leghorns) and lay a huge number of eggs, while others (Broilers) are ideal for food, but do not lay many eggs.

The next step is to pick the chicken breed you want.

Choosing the right chicken breed as a beginner can be the difference between thoroughly enjoying the moment with your chickens and wondering why you ever needed chickens in the first place.

Many people don't know exactly what they want or need until they start looking at the breeds, so that's why we decided to write this article today.

We will answer just what you need to think about, before you settle on your type, explain what traits beginners can look for, and then suggest our top 5 choices for the best egg-laying chicken breeds.

Why would you like chickens?

Before we go any further, we're going to ask you a few questions, be sure to write down the responses and keep them in your head.

First of all, why do you want chickens to be kept? Are you looking to keep chickens for eggs, meat, pets, or a combination of these reasons (though the meat and pet options probably don't go too well together!)?

Many chickens are exceptionally good egg-laying (Rhode Island Reds and Leghorns) and lay a huge amount of eggs, while others (Broilers) are ideal for food, but do not lay many eggs.

The second question you need to answer is, how much time do you expect to spend with, and take care of, your chickens? Many breeds require a lot of upkeep and effort from you. In comparison, other breeds (such as Buff Orpingtons) are very self-reliant and will not take much effort from you at all.

If you're curious how much time you need to spend with your ducks, read my project here.

Second, you need to care about your climate/weather to make sure it's ideal for the breed you're interested in.

Most of the time, you don't need to worry too much about it, because most breeds will be fine in all climates. Even if the chicken is delivered to you directly, unless the chicken has just been shipped, it will be perfect in your weather. However, if you're buying rare breeds (which we wouldn't recommend to you as a beginner, but more on that later) and you're traveling a long distance to get them, you need to make sure that the climate you're taking is right. Minorca chickens, for example, require very warm climates, so they would not be suitable in certain areas of Russia.

The fourth question you need to think about is how much room do you have for your chickens? Until you respond, make sure you read how much space chicken wants.

Were you planning to keep your chickens in a coop or in a free-range?

Some breeds need more room than others, and if they don't get this room, they can get violent and even start pecking and attacking each other. So make sure you match the breed to what you can offer in terms of roaming space and coop size.

The final question you need to talk about is what your plan is?

Some breeds pay a similar amount per chicken, but more exotic and uncommon varieties can be very pricey, they can potentially cost thousands of dollars.

Typical Chicken Characteristics Beginners Look For

We've noticed that most of the time, when people contact us wondering what kind of chicken they're supposed to start with, they're all looking for the same thing.

Most beginners are looking for chickens that are easy to keep, lay lots of eggs, are docile, and are not very noisy.

That's why we always recommend starting with what is known as dual-purpose animals. Dual-purpose birds are usually large egg layers, and very calm, we'll discuss specific breeds later on.

Many beginners contact us and inquire about rare breeds and breeds that produce a lot of food. We're not recommending either of these to beginners mainly because they need a lot more time, and they're harder to look after. We also recommend avoiding food or exotic birds until you have more practice.

BEST EGG LAYING CHICKEN BREEDS

So, if you're like us, and you want to start keeping chickens for eggs, which breed would we suggest?

Bear in mind that the ideas below are ideal for people with little experience looking for easy-to-handle backyard chickens, needing little upkeep, and, most importantly, laying a lot of eggs!

1. Rhode Island Red: Rhode Island Reds are associated with chicken keeping in the backyard and one of the most common chicken breeds in the country.

They're friendly, easy to keep, and very tough.

Eggs: should produce more than 250 medium-sized brown eggs per year.

Character: It's very easy to keep, it doesn't take too much room, and it's all year round.

2. Animal dogs, such as Golden Comets, have been bred to eat small amounts of food and lay as many eggs as possible. While this is good for you, it can hurt the hen's wellbeing because their body rarely rests.

Eggs: upwards of 280, medium-sized, brown eggs per year.

Character: Hybrids tend to make excellent surfaces, eat less meat, and are not likely to be broody. We make a great selection, just make sure you get your cross from a responsible breeder and make sure it's not over-bred.

3. Buff Orpington is one of the simplest and most common egg-laying chickens in the world. They reside from Kent, England, and are known for their good looks and sturdiness.

Eggs: should be produced at least 180 medium-sized, light brown eggs per year.

Character: Orpington makes great pets, as they are extremely friendly and soft. However, they do get broody during the summer months, so their egg production is slightly lower than some of the other breeds mentioned here.

4. Plymouth Rock, also known as barred rocks, originates from the USA and is one of the most popular dual-purpose chickens.

Eggs: are expected to produce 200 medium-sized brown eggs per year, which are also laid during the winter.

Character: A very friendly bird that performs best as a free-range and makes a perfect backyard hen. They're also extremely friendly to people, so good that you want to teach them to feed from your side!

5. Leghorn The Leghorn breed originated from Italy and was first imported to the USA in the 1800s. They don't get broody sometimes, and they're the perfect choice for year-round egg-laying.

Eggs: should produce more than 250 medium-sized white eggs per year.

Character: Leghorns will be great in the gardens as they are a very healthy bird, but they are not very tame, so they are not ideal for people with children who want them as a shelter.

With these recommendations, it is important to remember that you always get bad-chickens, and even the most docile breed will sometimes produce troublesome birds.

All of these breeds above should be available from the local hatchery, and at the beginning, we would recommend that you do not mix breeds within your flock.

SURPRISING RULES FOR FEEDING CHICKENS

Feeding your chickens is one of the most important tasks, if not the most important task when it comes to raising chickens in the backyard. Get it wrong, and you'll have a safe flock that is ready to cluck every time you bring them one of their favorite snacks or kitchen scraps!

Get it right, and it can lead to reduced egg production, deformed eggs, feather harvesting, and other undesirable behavior.

Now, let's get right into all you need to know about feeding chickens.

What Should You Feed Chickens?

Once you know what you're doing, feeding your chickens is going straight ahead. We think what makes this tricky are some of the unfounded assumptions posted online about what you can and can't feed your chickens (such as feeding your chickens potato skin is terrible for them, that's right! Chickens love potato skin).

A high-quality poultry pellet is the basis of any good chicken diet.

We feed our girls layers of pellets that give them the perfect amount of protein and minerals to stop them from laying eggs! Pellets usually contain wheat, salt, maize, sunflower seed, and oats.

Feeding your chickens pellets means that essential vitamins, nutrients, and minerals enter their food source to keep them healthy. This is even more important if your girls don't have much outdoor space because they're not going to be able to get minerals and salt from the ground.

In contrast to their simple pellet diet, you can feed them grains such as corn and wheat to give them some kind of choice.

Chickens like fruit and vegetables, and you can bring them this every day. The girls love potato peels, bananas, tomatoes, carrots, and broccoli. You are safe to feed chickens almost any vegetable or fruit except raw green peels (such as green potato peel) and any citric fruits such as oranges and lemons.

Only note that they need whole wheat, low salt, and low sugar meat.

Hens are healthy, and eggs are laid. We listed some of our favorite feeds in the table below.

How to Feed chickens

So now you know what you're supposed to feed your birds, the next question is how you're going to feed them?

Once in the morning and once in the evening, we feed our chickens pellets, remember that they like to eat small portions, but often.

Most people prefer to throw chicken pellets down on the ground and let their chickens hold to it. We put our pellets in a chicken trough to keep them clean and dry.

How much are you supposed to feed them?

Generally, free-range chickens won't eat anymore, so you can't force them. If you put too many pellets in their feeder, they just won't eat them.

Be careful not to leave any pellets or feed overnight, because it will attract pests such as mice.

Through time, you'll learn exactly how much feed your chickens require, and this will depend on the breed, how busy they are, and the time of year. If you are constantly finding feed in the trough, reduce the amount you give them slightly.

We've got 12 models, and we found out that every morning and evening, four-wide handcuffs keep them happy.

Interesting side notes: A hen requires approximately 4 pounds of chicken feed to produce 12 eggs (source).

How often are you supposed to feed them?

This is going to depend more on the situation than on the chickens. If you're unemployed and spend most of your time at home, you can feed them many times during the day.

Though, if you're busy and away from home all day, you're better to feed them once in the morning and then again in the evening when you're back home.

Another thing to keep an eye on while you're feeding them is to make sure the most aggressive (remember our pecking order discussion?) hens don't eat all the meat. If this is becoming an issue, consider feeding the weaker birds on their own to make sure they get some food.

Water for Your Hens

Providing your hens with water is very straight ahead, you just need to make sure they have access to clean, fresh water at all times.

You can put the water in any kind of plastic container, but the easiest way is to buy a drinker.

During the winter, when you stay in a colder climate, the water is likely to freeze in the evenings, so make sure you break the ice and wash the tank in the mornings.

Feeding Chickens Table Scraps

Of example, no chicken feeding conversation is complete without thinking about table/kitchen scraps.

One of the many benefits of keeping chickens is that the vast majority of your kitchen waste can be cooked. It means they're going to get a varied diet and you're going to save some money!

Make sure you try and feed your chickens with healthy foods such as rice, pasta, oats, fruit, vegetables, and wholemeal bread. As a general rule, if you can enjoy meat as much as they can. However, this excludes any fatty food or food with a lot of salt in it.

If we feed our girls scraps, we tend to cut them into little (thumbnail-sized) bits and chuck them straight to the floor in their cage. We're just putting the pellets in their trough.

You'd be amazed at some of the scraps that your chickens eat-pizza, spaghetti, and porridge, to name a few!

Before you feed your chickens' kitchen scraps, be sure to check your local regulations as in certain places (such as the UK). This can be surprisingly illegal.

5 Healthy Treats

Yeah, these chickens are probably spoiled! On top of their pellets and kitchen scraps, we're shocked that they still want to eat, but they do. Here are our top five healthy treats, which we occasionally spoil with:

Worms: they absolutely love worms.

Pumpkin: It contains a pumpkin plant.

Apple Cores: Simply drop the apple cores straight into the paper.

Broccoli: They can't get enough of it for some reason!

Porridge: They're just having this in the winter months though.

What you Shouldn't Feed Them

We've covered a lot of food throughout the post that you shouldn't feed chickens, so we're not going to repeat them again.

Certain foods that should not be fed to chickens include: avocado, rhubarb, garlic, chocolate, and any heavily processed meat (i.e., crisps).

Just remember, as a general rule, if you can eat chickens like that.

What Can Happen if Their Diet Isn't Right?

A great email we received from a reader last week was, how can I tell if my chicken's diet isn't right?'

The first thing to say was, should you note a significant change in their eating habits, make sure to have a vet look at them as soon as possible. And, if their food isn't perfect, there will be signs like that.

Egg production: If the season hasn't changed or their egg production drops drastically, this could mean that something is wrong in their diet.

General discontent and feather-picking: again, if the season hasn't shifted and they grab their feathers or each other's, this might mean that their diet isn't wrong.

Abnormal eggs: if you find that the eggs they lay are too small or consistently contain double yolks, this would indicate that their diet is not correct.

If you're searching for a helpful cheat sheet, be sure to check out what was developed by the Australian Government's Agricultural Department.

TOP 13 BEST EGG LAYING CHICKEN BREEDS

The Food and Agriculture Organization of the United Nations (FAO) has reported that in 2015 the cost of world egg production was USD 55 billion, and egg production was USD 70.4 million.

Before the Second World War, the largest egg production was obtained from small farms with around 400 chickens, but today, in large American egg-producing countries, chicken farms have reached 100,000 and even 1 million chickens in some cases. Nevertheless, given that the market for eggs raised in smaller chicken farms has increased, due to concerns about health and safety conditions in larger farms, the egg-producing sector has become very attractive to small enterprises.

Chickens are typically bred for meat or eggs, and each chicken farm must decide its business intent raising eggs, growing chickens for the meat industry, and providing a mixed farm, with both eggs laid chickens and chickens more appropriate for meat processing. As far as the most competitive egg-laying chickens are concerned, these are the best chicken breeds for this reason.

1. Australorp chicken breed

This breed, which originated in Australia, was produced in 1920, with deep roots in the Orpington breed. The name of this breed was given to the Orpington breed "Austral Orpington Club."

It's a chicken breed specialized in egg production, considered a true champion, as one hen of this type laid 364 eggs in 365 days. The Australorp breed is found in three color varieties: red, white, and black.

Australorp chickens are very healthy, with a rapid growth speed, beginning to lay eggs in the 5th month. They are ideal for growing in an enclosed environment but provide better egg production if they can walk freely in the open space. Such chickens are not good at flying, making it possible to expand in an open space, not having a large shelter. The hens of this breed are quite resilient, quickly going through cold winters, without impacting their egg-laying speed.

2. Lohmann Brown Classic chicken breed

It is the most common breed of laying chicken in the world and is used in almost every part of the world.

The Lohmann breed has a small size, with a bodyweight not exceeding 2 kilograms. A chicken of this breed produces up to 313 eggs per year, with a low feed consumption of just 110 grams per day.

3. Rhode Island Red chicken breed

The breed originated in the USA, where it is used for a dual purpose, both for eggs and meat. Such chickens are most common with small chicken farmers because they can easily adapt to backyard environments, have a high resistance to disease, and typically have a rather rough temperament. Rhode Island Red chicken can lay up to 260 eggs a year.

4. Sussex chicken breeds Similar to Rhode Island Red

This is a dual-purpose breed, which means that chickens are grown for eggs or meat. The breed has eight different colors, but the most famous are white chickens with black neck and tail feathers. Sussex chickens are very calm and tame, suitable for growing even in the yard. Chickens are capable of producing up to 250 eggs per year.

5. Golden Comet chicken breed

This breed is actually a widespread hybrid, known for its ability to produce between 250 and 300 eggs per year. The Golden Comets are very quiet and relaxed birds, ideal for open spaces and easy-going with other farm animals. The eggs have a brown shell on them.

6. Leghorn chicken breed

The breed originates in the port of Livorno, Italy, from a very old Italian community. It has 12 color variations, but the white chickens typically have the most feathers. Leghorn chickens have an average annual production of approximately 200 eggs, but up to 280 eggs may be produced. The eggs are 55-67 grams in weight and have a white shell. The brooding instinct is manifested very poorly, with the hens of this race brooding only in the proportion of 1-4 percent.

7. Marans chicken breed

It's a chicken breed native to France, with a very rich and colorful plumage. It has an annual production of 180-220 eggs and can be produced for both meat and eggs. The average egg is 60 grams, and the shell is brown.

8. Plymouth Rock chicken breed

This chicken breed is excellent for those who do not have a lot of experience growing chickens because they can easily adapt to a free-range lifestyle. These chickens are usually very tame and lay around 200 eggs a year, small to medium in size and brown in color. They can lay up to 280 eggs per year with proper care. The Plymouth Rock chickens are gray with white stripes.

9. Barnevelder chicken breed

This chicken breed is an unusual hybrid between the Asian jungle birds and the Dutch Landrace. Originally developed in Holland, it is known for its glossy feathers. The chickens lay up to 200 eggs a year, with a light brown speckled appearance and a small to medium size.

10. Buff Orpington chicken breed

Originally from Kent, England, this is the most popular variation of the Orpington breed and is ideal for backyard breeding because it is very tame and eager to socialize. Chickens prefer to get broody during the summer months, which is why they lay about 180 eggs each year.

11. Ameraucana chicken breed

This breed originated in South America and was named after the Araucanian Indians in Chile. It is also known by the name of Araucana. The Ameraucana breed developed as a hybrid between several hens and wild birds, from which they inherited the character that gives them the title of chickens that lay the healthiest eggs. The color of the eggshell has a blue-green shade; therefore, sometimes, it is also called "Easter eggs." It has an annual production of 170-180 eggs, weighing 53-60 grams. The egg production levels may not be important, but due to the low cholesterol content of the eggs, they are usually sold at a higher price.

12. La Bresse chicken breed

It is one of the oldest breeds of French chicken, created by a selection of local chicken populations. It has three color varieties: black, white, and gray, the most common being white.

A chicken produces 160-180 eggs per year, with a weight of 65-70 grams and a white shell. La Bresse is not only a good breeding chicken; it also has very tasty meat.

WHAT TIME DO HENS USUALLY LAY?

Raising chickens in the backyard is a fun and rewarding undertaking. Laying hens, bred for egg production, are plucky pets that reward your tender care with a steady supply of fresh and healthy food. Hens lay eggs during the day, most often in the night. The timing of oviposition or egg-laying varies with the breed of chicken and how much light exposure it gets.

Light exposure

The reproductive cycle of hens is controlled by photoperiod or light exposure. Hens need a minimum of 14 hours of light per day to lay eggs. They produce eggs at a maximum rate of 16 hours of light exposure. For particular, hens lay eggs within six hours of sunrise, or six hours of artificial light penetration for hens kept indoors. Hens with no access to artificial lighting in the hen-house should stop laying eggs in late fall for about two months. We start to lay down again as the days go on.

How the egg is made

Oviposition, or egg-laying, begins with ovulation. Hen ovulates by releasing an ovum, or egg yolk, from its ovary. Slowly, it travels down the hen's long oviduct, where the egg white, the shell layer, and the egg white form around the yolk. She lays the egg out of her cloaca, a common opening for her digestive, urinary, and intestinal tract. It takes about 26 hours, from ovulation to ovation.

Ovulation time

Hens usually ovulate in the evening but can ovulate as late as 3 p.m. The ovulation occurs about an hour after the egg has been laid. If a hen lays an egg in the morning, the egg may be postponed until the next day. The egg is going to be laid about 26 hours later. This is why hens regularly miss an egg-laying day.

Time of laying

A laying hen can at most only produce one egg every 28 hours. It does not ovulate and lay eggs in the dark, and the rate of egg production varies with the length and duration of the photoperiod. All these factors have an impact on the time of the day the egg is laid so that it varies with each egg. But it usually happens in the evening or in the early afternoon.

Genetic Factors

The hen's mating effects when the egg is laid. Brown-egg breeds tend to lie early in the day, while white-and tinted-egg breeds prefer to lie later in the day. Broiler hens, raised for meat production, lay eggs less often than laying hens. Broiler hens are allowed to lay their eggs before dawn.

HOW COLD IS TOO COLD FOR BACKYARD CHICKENS

How hot is it for chickens in the backyard? Below freezing, below zero, when it's too warm for me, there are all the responses I've learned. New chicken owners spending their first winter of chickens in a backyard coop may be worried about the actual winter weather. I live in a fairly mild region where the winter is typically not too extreme. We do get some winters that are colder than normal, and this often raises questions about how to keep the chickens happy and healthy. During the winter, I'm not too worried about my healthy flock members. Chickens are much cooler than heat tolerant. Indeed, changes in weather can have an effect on the flock. But if they're safe and well cared for, you shouldn't have chickens starving from the cold.

Signs of Cold Weather Stress

A chicken that feels stressed out of the cold is going to look cold. It might be huddled, not running around a bit. The feathers may be fluffed up considerably, and the chicken may sit on one foot, having the other tucked up in the feathers of the belly for warmth. This is the time to take action. Chickens are designed to self-regulate body temperature with their downy undercoat and increase food intake during cold weather. But there's a stage where they're going to need shelter even during the day. Heavy snow, rain, and below-freezing temperatures will need a few improvements to keep the flock healthy and happy. Don't wait too long, wondering how cold the chickens are.

A Well Designed Coop

A coop built for weather patterns in your area is the most important step you can take to plan for keeping chickens in cold weather. The coop can be heated before building, or the insulation can be applied afterward. Wrapping the coop in Tyvek's home wrap, or even attaching the tarps to the outside to break the wind, will help. The trick is to keep the wind cold and the rain out of the coop but to allow the water vapor to ventilate. Ventilation is so important to me. Don't keep the coop air-tight. Allowing good ventilation while still having warmth is crucial and will keep the chickens' warm while in the coop during extremely cold weather.

Holding the inside clean also avoids frostbites on combs, wattles, and feet. Applying a salve such as Waxlene to the combs and wattles will keep the frostbite from forming. Ammonia build-up will also be managed by keeping the coop clean. Clean up any drops easily. Imagine being in the coop during an ammonia smell outbreak! Not very friendly to any human person.

Should You Insulate the Coop?

A good coop structure will provide shelter from the wind, wet weather, and drafts. Isolation can be applied to the inside or outside of the coop. Hay balls are often used for insulation and can be stacked against the walls inside or outside the coop. Pay special attention to the north and west sides of the building.

Building insulation can also be applied during the coop construction process. Building a double-layer wall that traps air between the layers is one method. One is the use of traditional plywood-covered insulation to prevent chickens from pecking while filled.

I'm not in favor of adding mechanical heating equipment and heat lamps to the coop. Keep in mind that there are risks associated with the use of a heat lamp in a straw-filled system. When you plan on doing so, keep the 18 lamps away from any flammable material. Don't hang the lamp on the cord, and check the lamp frequently. I hesitate to discuss it because every year, a lot of farm families lose their whole flock and coop to a fire started by hanging a heat lamp in the coop. We have a regular bulb in our coop as a convenience to count chickens before we lock it up at night.

What About Sick or Injured Chickens and the Cold Weather?

It could be the best idea to move the vulnerable members of the flock to a cooler location for the duration of a cold snap. Weakened birds need TLC to maintain their strength and begin to heal. Using a crate in the laundry area and garage could be cooler to stop the chicken from using so much energy to keep it hot. During a cold snap, I brought a sick bird home to keep an eye on it. I'm sure many others have done the same thing. When you transfer the chicken back to the coop, once it's all right, do it on a warmer day. Being acclimated to rising conditions will be better if the day is colder.

Keeping the Drinking Water from Freezing

This is a matter for many of us in the winter. If you're not home enough to bring thawed water to the flock during the day, they won't be able to drink it when the water freezes. There are a few tricks that you can learn to use.

Using a small utility bucket, fill at least halfway with straw and sawdust. This will act as an insulating barrier and keep the water from freezing for a longer time. The air is also filtered by placing the bucket or rubber feed tub in the old tire.

Use a warmed dog water bowl during the day if power is available near the coop. Take it out and unplug at night so that the tank can be quickly refilled in the morning. For outdoor water containers, putting ping pong balls in the water helps the breeze keep the water from freezing. The movement of the ping pong balls will prevent the formation of the ice.

How Cold is Too Cold for Backyard Chickens?

It's hard to know exactly how cold the chickens in the backyard are. I believe it has a lot to do with the combination of proper housing and the overall health and fitness of your flock. There are people who raise chickens in extreme cold, without using artificial power. I'm surprised at the amount of warmth inside the coop on cold days, just because the chickens are inside.

For those of you who have harsh winter weather to months at a time, try these simple methods to keep your flock healthy and happy.   Be alert and watch the flock members closely, search for frostbite or symptoms of cold weather pressure.

THE MOST RARE CHICKEN BREEDS

As we try to make our way of life greener, we look to try and buy more organic vegetables, we've switched from plastic straws to cardboard ones. A lot of people look at having backyard chickens as a way to make their own eggs, maybe food or natural fertilizer. That, in effect, makes it possible to cultivate an organic garden and so on.

But when we try to save our world, we should also look at what we can do for the wonderful animals that are dying out. We may not be able to raise a black panther, but we might try to save a few heritages and endangered/rare chicken breeds!

Dong Tao

The Dong Tao chicken breed is very rare and more so. They're a breed that's very popular for its meat in Vietnam. Originally from the village of Dong Tao in the Khoai Chau district of Vietnam, these chickens are also referred to as Dragon Chicken. The males can get around 13.5 lbs. And the females are around 9.9 lbs.

We are called Dragon Chickens because of their extremely thick bodies and large feet. The arms could be about the size of the adult male's hands. Because they've got very big legs and feet, they're not very good with their own nests, not that a lot of hens lie down. We were cheating around 2 or 3 a week, and most of them were split because of the chicken's shaky big legs.

Breeders need to be quite vigilant in finding the eggs and getting them into the incubators so that they have a better chance of getting the hatchlings. They're terrible daughters, too! It's no surprise, though, that this breed is very rare and considered a delicacy in Vietnam. You're looking at about $2,500 for a breeding pair of this chicken breed.

Ayam Cemani

If you think the Dong Tao is expensive, wait until you see the cost for the AyamCemani chicken breed. They say these are the Lamborghini of chickens, they are rare, beautiful, mysterious, and they were actually considered a sacred bird in Java, where they originated.

Such perfectly inked black chickens have black wings, wattles, ears, comb, beak, and feet. They hold on to themselves in such a way that a person can actually become quite mesmerized by this truly unique bird.

Another very special thing about this chicken breed is that it's not just its outer appearance that's completely black! The body, the liver, its hair, and even its bones are gray.

The only thing that isn't black (though it's really dark) is its water. It was also a very common bird for various rituals.

Now, this breed is very rare, and if you can get them, there is usually a waiting list for them up to a year in advance. They can also get a price of $2,500 each, which is $5,000 for a pair of them!

AyamCemani is a mythical bird from Indonesia that has been connected with many mysterious, healing, and lucky charm legends over the years.

If you see this truly gorgeous hen, you can understand why. AyamCemani stands proudly upright, wearing her beautiful, soft, glossy black feathers that shine in the sun with a beetle of green and purple. This is a truly unique bird, in that not only are its feathers black, but comb, wattles, earlobes, ears, skin, ribs, and even organs black. While their blood is the natural color, it is also very dark and has been used for religious, sacrificial and ritual purposes throughout the ages in Indonesia.

Because of this, the bird has been used for many ceremonial rituals and is held in the highest respects. It is one of the rarest breeds, with most breeders usually sold out for up to a year in advance. Not only is it rare, but it is also one of the most expensive breeds, which is why it is known as the Lamborghini of poultry.

In many areas in Indonesia, these birds are highly valued, as they are considered holy animals. They're supposed to bring luck to those who keep them, and there's a belief that eating their meat can cure certain diseases.

Being such a rare bird, it doesn't bode well for their numbers that they're very common for various rituals.

History

AyamCemani is from Jakarta, Indonesia. We were thought to have been born from the Kedu, which is believed to have originated from the AyamBekisar, this breed of chicken hails from a group of islands in Indonesia. The origins of the AyamBekisar are believed to be the combination of domesticated red jungle birds with green jungle birds.

In the past, and still today, this breed is used as a foghorn by some seafarers on their boats because of the very distinct and unusual breeds of the crow.

The name of the Ayam part of the chicken breed means chicken in Indonesian. Cemani can mean either solid black in Sanskrit or Cemani village in the local Indonesian dialect.

These birds have been used as a symbol of community standing and wealth status. They're not the common people's chicken.

Where To Find These Birds To Add To Your Flock

AyamCemani is not a chicken breed that you can easily find in any nearby poultry and ranch. You will need to order them from a licensed registered breeder with a strong bloodline stock. You will locate such breeders through the AyamCemani Breeders Association or on farms such as Greenfire Farms or Feathers Lovers Farms.

Check with your local animal welfare and shelter for more information on how to take the rescue, Plymouth Rock. They will also be able to help with any special needs, attention, or care they may need.

GENERAL

AyamCemani is an extremely rare exotic beauty with mystical qualities, the entrance of a person, and the personality that delights you. If you're hunting for the chicken that's going to be the star of your flock, that's the chicken for you. Although, you're likely going to have to wait up to a year to collect your bird(s). You can get them at certain lower end poultry vendors, but the quality and breeding requirements of these beauties will not be the same.

GROOMING

Make sure to de-worm your beautiful beauties on a regular basis, particularly if they are free-moving around the yard, playing with other domestic animals, and engaging with children. A good dust bath or two put under a comfortable shade tree or canopy with a touch of sage or lavender mixed in will be greatly enjoyed by these ladies. I love taking a dust bath to rid their beautiful feathers of insects and waste oils. A regular weekly or biweekly check for mites, lice, and various other parasites should be done to ensure a healthy flock.

DIET OR NUTRITION

These chickens are part of the occasional table scraps or consume most of the remaining vegetables or fruit as a tasty treat. As with other chickens, chicken pellets, grains, chicken mash or grain mixtures from 8 weeks of age and older are eaten. It's always best to feed your chickens their regular food first thing in the morning before they're allowed to forage.

The best thing for baby chickens is Chick Starter when they're under eight weeks old.

Laying hens should have extra protein and calcium in their diets to ensure the quality of their eggs to keep them in tip-top shape.

Please see our comprehensive guide to "Feeding your chickens" for more information on the different types of chicken feed for chicks, hens, laying hens, roosters, etc. and where to buy food and approximate feed costs.

SOCIALIZING THE BIRD

AyamCemani is mostly respectful of other breeds and their own animals. They have a calm nature, and they're going to settle in with just about any other breed. We make excellent mothers and brood chickens to feed other chicken eggs.

Always check how well the breed is going to get on with your current flock before you buy it, as you don't want to upset your coop or stress your current flock.

As with any newcomer to the roost, you will have to quarantine the bird for 7 –31 days to ensure that there are no unwanted critters or diseases that could spread to your current flock.

Onagadori

The Onagadori is a large bird that originates in Japan, where its name is loosely translated as "honorable bird." This breed was designated as a National Treasure in Japan in 1952, and to date, there are only about 250 birds left in Japan. These are exceptionally beautiful birds in which the tails of the roosters grow from a distance of 1.5 meters. One of the longest tails of This species recorded was 12 meters long. They've got about eighteen feathers in their tails that don't molt.

Their saddle hackles have also been known to be quite long and tend to grow quite fast. These come in three different color variants, black-brown yellow, black-brown gray, and green. The white range is my pick!

We need a considerable amount of attention and attention to their feathers. Most of the owners will keep them poised so that their feathers do not fall on the ground and cover their tails with silk ribbon to keep them in perfect condition.

Such magnificent ornamental chickens are raised as display birds and have been declared a Special Natural Monument in Kochi, Japan, where they originate.

He word Onagadori means eight "Long-tailed chickens" or "Honorable Fowls" in Japanese, who have had great pains in their breeding over the years.

The roosters have exceptionally long tail feathers, which do not molt so that they can grow too long for the whole life of the rooster. The minimum requirement of which their tails must be at least 2 meters long if they are to be considered a Longtail fowl.

The remainder of the feathers of the rooster, which cover the breast, the back, and the neck, usually molten like the feathers of the hens.

The Onagadori tail feathers have been recorded to be up to 27

There are other species, such as the Phoenix, which is a descendant of the Onagadori, which also have long tail feathers, but typically do not reach nearly 2 meters, and their tail feathers are liquid.

These beautiful birds require special housing, perches, and their tail feathers need extra care to be kept in top condition. Sometimes their tails are rolled up and secured with silk ties for protection, as the breeders take great pain not to allow the tail feathers to be dragged to the ground or to be roughed up, but to the wind and other elements.

HISTORY

The Onagadori is one of Japan's treasured monuments which, once declared as such, prohibited the export of eggs from the breeds. Therefore, every Onagadori seen these days outside Japan comes from eggs that landed in those countries before the bird was declared a Natural Japanese Monument.

It is because of this prohibition that these birds have been put as important on most preservation lists.

The roots of the Onagadori breed are quite ancient, and its ancestry is believed to have been a mixture of the green male jungle fowl and the red female jungle fowl. Such dogs were called Bekisar and originated from the Isle of Java.

Such jungle-birds were well known for their lovely long tails and crowns.

The hybrid of these two chickens was used to start a new breed of chicken, of which by 206 BC, there were many male birds with a long flowing tail.

These birds brought good money to Japan to export to China. Another breed came from these long-tailed chickens that were exported to China and called the Shokoku.

Approximately 700 to 900 AD, it was from the Shokoku chickens which Japan introduced from China that Japan bred another species called Totenko. The Totenko was brought up in Kochi and recorded by Professor Hiraoka.

It was from the crossbreeding of Totenko and Shokoku that the Onagadori was born, and the genus was improved from around 1912 to 1926 (Taiho period) to what we know today.

In the 1920s, to protect their lovely ears, they came to be bred as cage birds.

The first Onagadoris to arrive in Europe in the mid-1940s were known because Yokahama's as they left the port of Yokohama in Japan. These birds were not very adaptable to the European environment, prompting breeders to breed them with native, stronger breeds that shared the genes of Onagadori. As a result, a more robust Onagadori breed has come to be.

There are a few Onagadori in the USA, but they are very rare to come by, and they are often mistaken for the Phoenix Longtail.

Polverara

The Ploverara chicken breed is a very new one, others say, an antique breed that comes from Italy. They are a medium-sized crested chicken with a V-shaped red comb, small wattles, and white ear-lobes. They also have a beard that, together with their unique shaped crests, makes them look like little-bearded chefs. They come in two color variations, white and black. The white variant is pure white, and the black variety is the dark pigment that reflects a red beetle in the light.

Such chickens make incredible birds of prey. They lay a decent amount of medium to large white eggs. About 150 per year and their meat are little darker than most other chickens, but it is known for its delicious taste.

Ixworth

The Ixworth chicken breed was developed in Ixworth, Sussex, UK, and is becoming increasingly rare every year. This is sad because this dual-purpose chicken breed is an all-around, cheerful man and a great addition to any flock. It only comes in pure white and will lay between 160 and 220 cream / tinted eggs a year. Eggs are usually small to large in size. It was also a really good one after that for its very juicy, tasty meat. They're a large breed of chicken that's easy to handle, calm, and can be tamed.

Naked Neck

The Naked Neck has long been thought of as the "Turken" This is because it looks like a duck and a rabbit has been hit. However, there was once a story that this was the case. It is not a mutation that the chicken breed was designed to give the chicken up to 50% less feather protection than the normal chicken breed. Most of which are most visible on the chest where there are normally little to none. That's why these birds look like Turkey!

They come from Transylvania, which is why they are also known as Transylvanian Naked-Neck ducks. We have an outstanding quality of meat, which is very juicy, and are really good layers of large brown eggs. You can predict from between 200 to 250 eggs a year per hen.

Such chickens may look different, and at first sight, some people may even think there's something wrong with them. At first sight, you can see why people would think this way. The Naked-Neck chicken has up to fifty percent fewer feathers than other breeds, and a naked, plucked neck. That's why they're called Naked-Neck chickens or Transylvania Naked Neck chickens!

They may not look that pretty, but these birds are a very hardy breed that doesn't mind the weather, even with low feathers. They're going to need some nice warm shelter to warm up after a wintery walk around the garden. They can also withstand heat, but they will need some attention as they can get a sunburn, but there are easy preventive measures for this.

They're a big bird with a soft, friendly nature, and they can be quite amusing, as they're also quirky with their own amazing characters. They make great pets with a bonus of laying a decent number of medium-sized eggs for the table, and they are known as roasters for their good quality meat. They're actually quite a popular chicken in the backyard in France.

HISTORY

The Naked Neck chicken is believed to have originated in Transylvania. However, the breed as we know it today was mainly developed in Germany. The Naked Neck chicken was developed as a smooth-skinned dress bird that was much faster and easier to pluck.

The Naked Neck chicken was once believed to have been a hybrid cross between domestic turkey and chicken. As a result, the breed was often referred to as Turken. While the Naked-Neck chicken really looks like half a turkey and half a chicken. Scientist has long since proved that the two species are genetically incompatible. The bare neck trait in a chicken is caused by an "incompletely dominant allele." The Naked Neck gene has been shown to improve the size of the chicken breast and reduce the heat stress in some chicken breeds. If the gene is bred into different broiler breeds, it can help lower the chicken's body temperature to help with better body weight gain.

The Naked Neck chicken has proven to be an excellent forger and is also immune to most diseases.

They have been and still are a very popular bird in Europe, particularly in France and Germany.

However, they are not very popular in America, with very few breeders and backyard farms holding them. This is a pity because these chickens are very easy to tame, friendly and very calm. They were first seen in America around 1924 and were accepted into the Perfection Standard by the American Poultry Association in 1965.

Golden Campine

The numbers of the Golden Campines are in decline as they are usually overlooked in the case of breeds that are tougher to the climate, mature faster, and lay more eggs. But these smart, successful, and responsive birds need help to ensure that their numbers rise again. Just because they do not grow as quickly or almost as much as commercial/hybrid breeds does not mean that they are not useful. Hens lay about 180 to 200 medium white eggs a year, and if you kill chickens before 18 months, they make a nice bird of food. Not too big, but they're going to give a decent meal.

The breed is also unique in that the males are what is considered "hen feathered," which means that both males and females share healthy colors and patterns.

They're often considered to be an old breed that had spanned back to the time of Julius Caesar, who was believed to have owned a handful. To be more precise, he took a few to Belgium after a looting spree.

The Campine is a rare, uniquely beautiful chicken with a very upright stance, with tightly knit feathers that make them look larger than they are.

Such birds are great fliers, and they can be a bit sketchy and driving. They are very inquisitive birds, and although they don't like to be treated badly, they are very friendly. They're going to follow a human around the garden and talk back. They're not what's called a sitter, and they're very rarely going to crumble, so if you're trying to breed Campines, you might need a bridal hen to sit on their eggs.

Such busy chickens take quite a bit of space to roam around and forage about.

Where To Find These Birds To Add To Your Flock

Campine is a very rare bird on the vital conservation list, so it is best to find a certified breeder by organizations such as the Livestock Conservancy, the American Poultry Association, and the Society for the Conservation of Poultry Antiquities. There is also a club in the UK called the Rare Poultry Society, which includes breeders from all over the world, a lot of useful information and advice about the Campine. Try and avoid purchasing this rare ancient breed from poultry shops and plantations, as there are not many Campines in America. Registered breeders will also be able to help with any special needs, attention, or care they may need. If you're planning on breeding your chickens, you're going to want to make sure they're from a good bloodline.

CARING FOR THE BIRD(s)

This is a comprehensive guide to chicken ownership. It covers where to start by choosing your ideal flock, the coop that would best suit your garden, your bird, and you to buy and bring your bird(s) home.

GENERAL

An exceptionally rare beauty that will be the treasure and centerpiece of any flock with its unique plumage and coloring.

GROOMING

One or two well-positioned dust baths are going a long way to keeping Campines pest and oil-free. There are several supplements that can be added to the soil to ensure that their feathers stay perfectly good. This may be quite a challenge, as these chickens will skitter away from being picked up, but they will have to be inspected for mites, lice, and any other parasites at least once a week. Always get your birds de-wormed on a regular basis, especially if they are around other animals or interact with children.

DIET AND NUTRITION

Make sure you get your regular chicken pellets, grains, chicken mash or grain mixture from 8 weeks of age and older first thing in the morning. This must be fed to them before they go out to forage every day to ensure that they are well fed.

The best thing for baby chickens is Chick Starter when they're under eight weeks old.

Laying hens should have extra protein and calcium in their diets to ensure the quality of their eggs to keep them in tip-top shape.

Vegetable and fruit table scraps are always welcomed by these busy birds and can go a long way to get them to comply with their weekly pest control.

Feeding your chickens properly would give your organic garden a lot of nutritious fertilizer to help your vegetables and flowers develop.

Please see our comprehensive guide to "Feeding your chickens" for more information on the different types of chicken feed for chicks, hens, laying hens, roosters, etc. and where to buy food and approximate feed costs.

SOCIALIZING THE BIRD(s)

Campines tend to shy away from other breeds, and it takes a long time for them to trust newcomers. They may be a flying sketch bird, but they're not an overly aggressive chicken, so it's best to get chickens that have a calmer nature to mix with Campines. They're not at all bridal hens, so if breeding is intended, it might be wise to look into getting bridal hens to add to the flock.

Always check how well the breed is going to get on with your current flocks before you buying it, as you don't want to upset your coop or stress your current flock.

As with any newcomer to the roost, you will have to quarantine the bird for 7 –31 days to ensure that there are no unwanted critters or diseases that could spread to your current flock.

Vorwerk

The Vorwerk is a dual-purpose chicken breed with a slate gray skin color, a stunning plumage, and a wonderful personality! They're not quiet animals, so they want to know what's going on around them. They are great foragers, have great meat, and lay about 180 medium brown eggs a year. Originally from Germany, they were built by Oskar Vorwerk. Oskar wanted a chicken breed that was an all-around pleaser, One that could lay a decent amount of eggs, have a great quality of meat, and look good. The Vorwerk was developed by the Andalusian, Buff Orpington, Buff Sussex, and Lakenvelder breeds.

Modern Game

Early Game breeds are mostly kept as display or ornamental birds these days. A long-legged, straight-up stance is similar to the supermodels of the poultry exhibition runways. We even come in a variety of stunning plumage colors that gives them the advantage of most poultry displays. Unfortunately, this breed is not as common as it used to be, and their numbers are declining. These beautiful birds take some care of themselves, as they are not very cold hardy and tend to prefer the hottest climates. But they're friendly birds, and they're full of lovely antics.

Classic Game Chickens are display and ornamental birds that have been raised as stunning counterparts to their gamecock ancestry.

We are the perfect show chicken from their dubbed combs and wattles to their stunning, willow, and dark faces. They are a tall, slender chicken with long, clean legs, well-shaped toes, beautiful color, and erect postures that make them look like supermodels of chicken breeds.

Shells: These are very weak layers of shells.

We lay medium white eggs. They lay between 50 and 80 eggs per year. They lay throughout the year but are more likely to lay in warmer weather. They begin to lay eggs at 22 weeks old.

Meat: They've got white skin. They're not meat chickens Breeding: they're not too difficult to breed unless you're going to show them. If so first time/novice breeders should seek advice from more advanced registered breeders Hens do get broody They are not really good brood hens They will sit on their eggs They will raise their chicks making surprisingly good mothers They have been known to mother other chickens/breeds chicks Show Bird: They were bred to be a show bird their bodies, posture, and personality aesthetically. These are the supermodels of the Poultry Show Houses.

Pets: Modern Game is sweet and inquisitive, who are attached to their humans. Other: Although they will not lay a lot of eggs, they will grace your garden with their beauty and style. They're not scratching almost as much as other breeds, so your garden and lawn won't get too messy from them.

HISTORY

Modern Game hens were developed due to the prohibition of cockfighting in 1849. Owners of fighting cocks took to display their birds at poultry shows, creating a new use for their game chickens.

Over the years of displaying these game birds, the poultry expectations for them have taken on more form, and more attractive qualities have been found in these game birds so that birds can gain awards and win shows.

With the aim of obtaining these requirements and producing winning bird breeders, we have begun to develop what we know today as the Modern Game Chickens.

The breed was first referred to as the Exhibition Game in the late 1800s and was not nearly as tall as the modern-day Modern Game Chickens.

By the 1880s, there was a dispute that the chickens should not be as tall as they were, and that some of the breeds used the name of the Old English Match.

By 1910 the breed as we know it today was developed and named Modern Game. Their sleek, tight, feathered frames, long, elegant legs, and hard feathers were all priced. They had quickly become a very popular bird show, and the Modern Game Club had been set up in England.

During the First World War, their numbers declined rapidly, as they were not a very useful bird, and the money was tight, so their popularity dropped.

By the time of the Second World War, their numbers were so few that the breed was thought to have become extinct.

But fortunately, the breed had also made its way to America in the late 1800s, as well as to various European countries.

The breed was brought back to life in the 1960s by Paul Hohmann, a German Breeder, who managed to get some eggs from America.

The Bantam variety of the Modern Game was developed in the 1860s, and in the 1900s, the breed was developed into the variety available today. As they were much cheaper and less expensive to maintain than the large Modern Game variety, the number of Modern Game Bantam varieties did not decline during the First World War.

TAKING CARE OF YOUR CHICKEN FLOCK

Your birds will require quite a bit of your daily care and well-being.

Keep your bird(s) healthy.

Vaccinations

It is best to tell the manufacturer what birds have been vaccinated against or before you buy the chickens.

If your chickens are combined with other poultry and domestic animals, it is best to have them vaccinated for Marek's disease.

It is also advisable to check vaccinations for your bird(s) with your local vet, animal shelter, and or animal welfare society.

Prevention is always better than cure, so if you can stop an infection, it's best to do so, particularly if it's widespread in your community.

Chickens are also prone not to show that they're sick until it's too late. We must behave as normal as possible to avoid being bullied by the rest of the flock.

You can handle most of the vaccines yourself, but it's best to let a doctor or an animal husbandry specialist do it.

Make sure you keep the vaccines up-to-date.

De-worming

All animals kept domestically or on a farm should be dewormed at regular intervals, especially as these pests can spread at an alarmingly rapid rate to infect your flock.

Preventing / Treating worms: There are different methods to do this, but it's always best to check with your doctor for the right one for your flock.

We will also instruct you on how often how to prescribe the drug to them.

Over counter solutions: one of the easiest and most cost-effective ways, costing around $20, is a liquid wormer.

You put the amount you need in their drinking water about once a month.

Natural remedies: There are also more natural remedies that you can do throughout the month that will also help to get rid of any existing worms and even some other pests like sand mites and ticks.

You can try crushing some garlic with your food or giving the pumpkin seeds, and even Apple Cider vinegar is going to do the trick.

Severe cases: In severe cases, it is always best to take the animal to the vet and ensure that you get the proper treatment to treat the rest of your flock.

An outbreak of worms: if your flock is getting worms, it is best to ensure that any other animals around them are also dewormed.

Make sure the ground on which they walk, feed-in, etc. are also worm-free.

Surroundings: there are products that can sanitize the soil to ensure that the worms are killed.

Keeping the grass short will help the sun's rays to kill any remaining parasitic eggs that may still be in the grass.

There are chemicals such as Soubenol, Flubenvet, and other herbal treatments that can be obtained from your local vet or veterinary suppliers.

Symptoms of worms: It's not easy to tell if your bird(s) have worms, but in the early stages, they may leave to lay eggs in unusual places.

They may have diarrhea with some worms; you will be able to see them in the chicken's droppings

They may eat more

They may lose a lot of weight

And if they have a gapeworm, it will stretch their necks as if they are gasping for air

Severe infestations can be fatal to the bird(s)

Types of worms:

Tapeworm

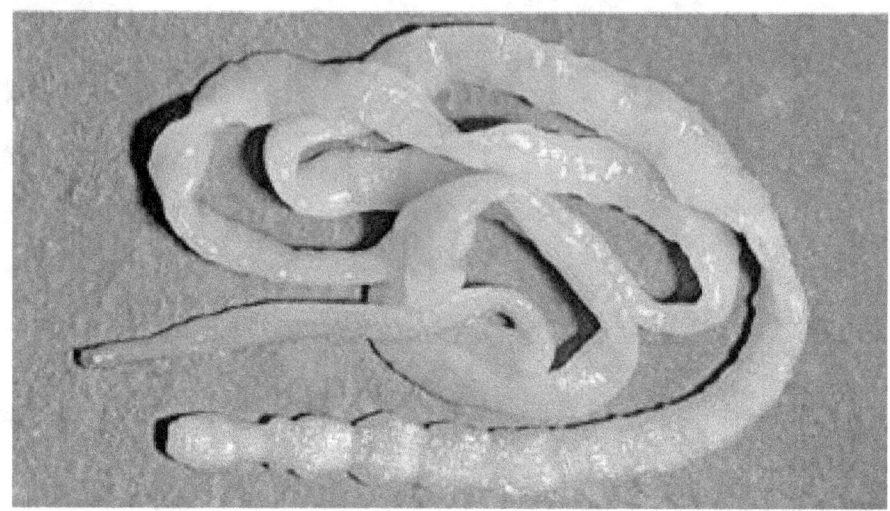

This is not that severe, but it can happen and damage the bird(s) immune system when they bind themselves to the intestinal lining of birds. You will need to take them to the vet and decide if they have tapeworms and the best treatment for this.

Gapeworm

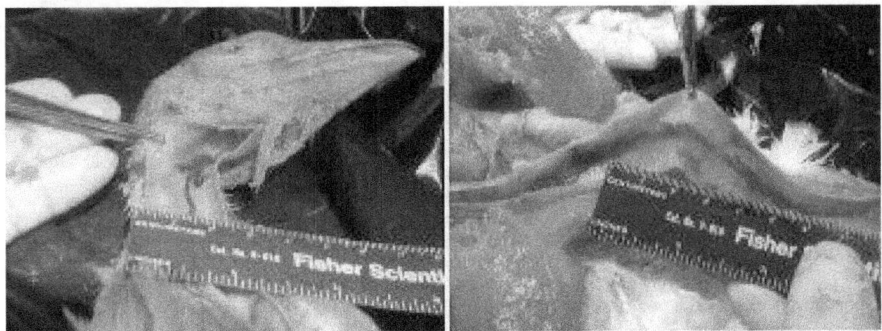

These nasty worms cling to the bird(s) trachea, and that's why they make the bird stretch out their necks and look like they're gasping for air. Such worms are more common in chickens as they can be picked up from hosts, including snails, slugs, and worms.

Round-Worm

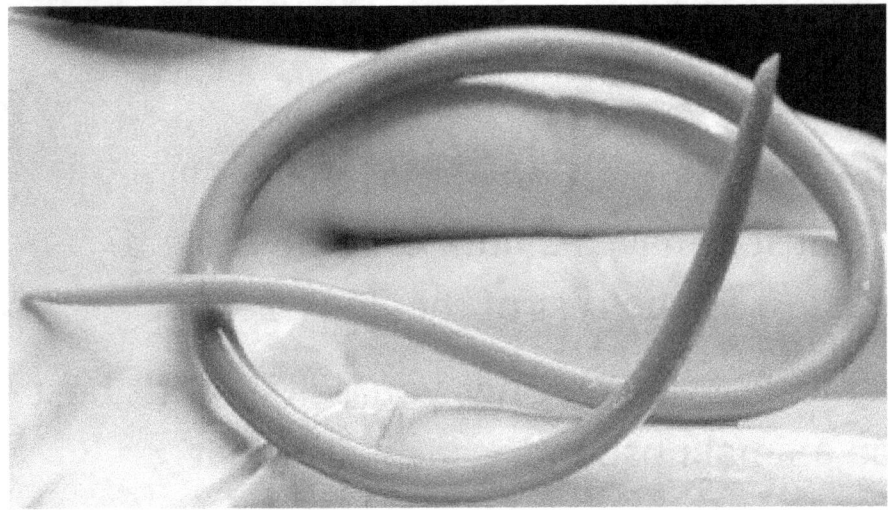

TAKING CARE OF YOUR FLOCK

Your birds can take quite a bit of your daily care and well-being.

Keeping your bird(s) on safe Vaccinations When purchasing your ducks, it is best to tell the manufacturer what the birds have been vaccinated for.

If your chickens are combined with other poultry and domestic animals, it is best to have them vaccinated for Marek's disease.

It is also recommended to confirm vaccines for your bird(s) with your nearest vet, animal shelter, and or animal welfare society.

Prevention is always better than cure, so if you can avoid an infection, it's best to do so, particularly if it's widespread in your town.

Chickens are also likely not to reveal that they're ill until it's too late. We must behave as normal as possible to avoid being harassed by the rest of the flock.

You can handle most of the vaccines yourself, but it's best to let a vet or an animal husbandry specialist do it.

Make sure you keep the vaccines up-to-date.

De-worming All animals kept indoors or on a ranch should be dewormed at regular intervals, especially as these pests can spread at an alarmingly rapid rate to kill your flock.

Preventing / Treating worms: There are different methods to do this, but it's always best to check with your doctor for the right one for your flock.

We will also instruct you on how often to prescribe the drug to them.

Over counter solutions: one of the simplest and most cost-effective ways, running about $20, is a fluid wormer.

You put the amount you need in their drinking water about once a month.

Natural remedies: There are also more natural remedies that you can do throughout the month that will also help to get rid of any potential worms and even some other insects like sand mites and ticks.

You can try grinding the garlic with your meat and offering the pumpkin seeds, and even Apple Cider vinegar is going to do the trick.

Serious cases: In severe cases, it is always best to take the dog to the vet to ensure that you get the proper treatment to treat the rest of your family.

An epidemic of worms: if your flock has worms, it is best to ensure that any other animals around them are still dewormed.

Make sure the soil on which they run, eat on, etc. are also worm-free.

Surroundings: there are materials that can sanitize the soil and ensure that the worms are destroyed.

Holding the grass short will help the sun's rays consume any remaining insect eggs that may still be in the grass.

There are drugs such as Soubenol, Flubenvet, and other herbal remedies that can be purchased from your vet and pharmaceutical distributors.

Symptoms of worms: It's not easy to tell if your bird(s) has worms, but in the early stages, they can leave to lay eggs in unusual places.

They may have diarrhea—with some worms you will be able to see in the chickens falls, and they will feed more, they will lose a lot of weight And if they have gapeworm, they may extend their necks as if they were gasping for air Severe infestations can be deadly to the bird(s) Forms of worms: Tapeworm is not that widespread but can happen and will damage the bird(s) immune system when they strike.

Gapeworm Photo Source: extension.umaine.edu These disgusting worms are bound to the trachea bird(s), and that's why they make the bird stick out their necks and appear like they're gasping for air. Such worms are more common in chickens as they can be picked up from hosts, including snails, slugs, and worms.

Roundworm, there are a few types of these worms, the most common of which is the big roundworm. These can basically take place anywhere in the digestive system of the bird(s). You may be able to pick them up in your bird's drops.

Pest-free There are other forms of pests other than worms that can affect your flock.

These are quite normal and may happen even with the regular care of your coop/run. Fleas enjoy the warm weather and can be quite a nuisance in the summertime, particularly in the hottest climates.

Prevention / Treatment of fleas, mites, and lice: cleaning and treatment of coop, run, garden areas, and other domestic animals that may come into contact regularly is a must.

Check your birds often by inspecting their feathers close to their skin, particularly on their breasts, tails, and vents.

Several mites are going to get between the scaly areas of the chicken legs.

If they have one of these infections, they can be handled with different solutions depending on the parasite.

Over the counter solutions: you can get any supermarket or pet hospital over the counter treatments.

Please read and follow the guidelines if you are not certain that you will seek advice from your nearest vet or animal shelter.

Ensure that the areas in and around the coop or wherever the chicken's play has been cleaned.

Natural remedies: There are garlic sprays available, or a person can make their own by smashing 3x the whole cloves of garlic in 2 cups of water. You can also add a teaspoon of lavender, spearmint, thyme, berry, or cinnamon oil to the mixture.

Spray chickens, coop, walk, and a number of locations around the coop.

You can even add crushed garlic to their water or dry food.

Extreme cases: In severe cases, it is always best to take certain types of flea from the animal to stay near the eyes of the bird, and a serious infestation of these can lead to blindness. If the chicken is in severe discomfort, it is best to have the infestation treated by a vet.

An insect outbreak: When one bird has fleas, it is almost certain that the rest of your flock has fleas. Every bird must be handled, its coop fully washed, just like the surrounding areas.

Ensure that all the animals are handled as well.

Once you've washed the coop, make sure you take off any socks, boots, etc. before you get back. These should be washed and cleaned up afterward, as you don't want to catch them in your house.

The surroundings:: make sure you keep the grass low as these insects appear to grow in the lawn.

There are both chemical and natural treatments in the garden to get rid of the pests.

Symptoms of pests such as mites, fleas, and lice, these three pests tend to show the same signs of infestation. Your birds will show the following most common symptoms that may be irritated by these parasites: discomfort and blood loss Decrease egg production Continuous scratching Irritability General malaise Restlessness See feathers deteriorate in patches and have raw skin spots Some infestations may cause blindness and even death. The pests that you may encounter with your bird(s) are:

Mites:

Such scary little red critters are distantly related to no-other than a spider.

They do come in three varieties:

Northern Fowl Mite: they love to live in the Coop's vent area, but they are quite happy to live on chicken as well.

Red Mite: these are the most annoying mites, and once an infestation happens, it's really hard to get rid of them. You may even need to get a qualified pest control to get rid of them. We love to live in nest boxes and other coop nooks and crannies. They're going to feed on chicken's blood during the night.

Scaly-leg Mite: these horrid little critters live in the scaly areas of the chicken leg. This can make the bird go lame if it's not handled on time. They're going to make the chicken leg thick, rough and swollen.

Lice:

Such pesky critters suck on the feathers of the chicken and the flakes of the hair. They are very annoying to the chicken, and they lay their eggs at the base of the feathers. Several situations are so serious that the bird is unable to collect the eggs.

There are about 50 different kinds of lice that can be found on chickens. Ask your local poultry specialist for more in-depth information on these.

Fleas

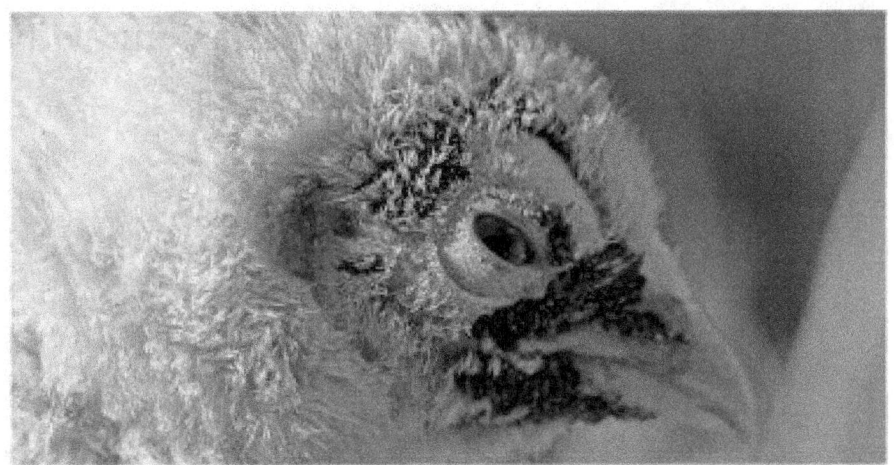

Most of the chicken fleas you can see are only about 5% of what is actually on the bird. The remaining 95% or the flea species is either larvae or egg-shaped on the animal.

Fleas can survive in a long grass for a short time, but they need a host to survive. They live on a bird and tend to breed rather quickly as soon as they are well fed.

Consider sprinkling FGDE on the coop wall, nesting boxes, roosting areas, and it can also be dusted on your flock (always check the directions before applying it).

While garlic is excellent as a general deterrent to most pests and worms, there is a risk of Heinz Anemia. This is because garlic is part of the family of onions. Onion is not a recommended source of food for your birds.

Apple Cider Vinegar sprinkled gently on your flock feathers as well as some herbs like mint will keep them away.

Ticks

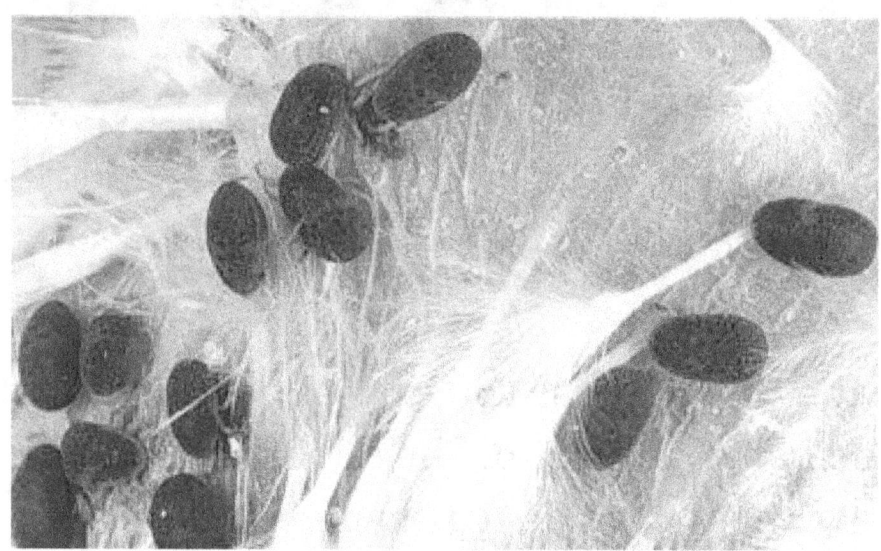

Poultry ticks are sly critters that you may not notice on the bird. That's because they're hidden in the coop during the day to feed their host at night.

Such ticks are also borne by wild birds and are very common in nests such as vultures, buzzards, egrets, etc. They live in dark crevices between the wooden slats of the coop or in cracks in the floor.

They love the heat, the damp, and the humid air. Although the very cold weather is going to kill them, it's best to keep your coop tick-free no matter the season.

The creepy thing about poultry ticks is that they can survive without feeding for up to 4 years.

Effect on the flock: There are many serious diseases that these critters can inflict, including increasing paralysis of the foot, wing, and respiratory process. They can cause severe anemia, depression, and weight loss to the bird.

Signs and symptoms of chicken ticks: Agitated chickens at night, they can hurry and do not want to coop, they may become aggressive, and general malaise in the bird(s).

Increased egg production Treatment: There are alternative treatments that can be obtained at the local animal veterinarian or live poultry distributor.

If the chickens are housed in your local animal shelter, they will be able to help you with the cure.

Because ticks may have some nasty side effects if you've had a Poultry tick epidemic, it's best to get your flock /bird(s) checked over by the vet.

De-tick the coop Clear the birds from the coop.

It may be best practice to set up a clean, temporary home for a day or two.

Remove all the nesting boxes, the water, the food bowls, and, if you can, the roosting pots.

Use the cleaner that you purchased and clean the coop, nesting boxes, and roosting posts with it.

Use a high-pressure hose to blow out coops, nest boxes, and clean the roosting posts.

If there are holes between the frames, it may be safer to cover them with caulk once the coop has dried out.

If your nesting boxes are also made of wooden boards that have holes, caulk them, too.

Such ticks like to live and breed between sheets, knots in the wood, or small cracks or crevices they may locate.

Once the coop has been cleaned, you can let your birds go back to the coop.

Once you've had one infestation, it's best to always check it out every time you clean your coop.

Grooming Although most chickens do not require a lot of grooming, they still like to keep their feathers well trimmed and clean.

Show birds do require extra care and maintenance, even a bath! The daily gals can also be bathed periodically.

Bathing your chickens

Sometimes you may need to bathe your hen(s) like most other pets.

You're going to need a big bath to dip the bird in comfortably.

Fill the tub with warm water. You can put some soap sweaters in your bathtub, some baby shampoo or some mild soap.

Or you can gently apply the soap to your feathers.

Ensure that you hold her softly but firmly and that you don't worry too much about her.

Once you've soaped her, make sure you wash all the soap off her feathers.

Several people are going out and giving their hens a good dry blow with a hot to a cold hairdryer.

Clipping nails

Most of the chicken's nails are obviously damaged from their foraging and scratching. But there are some occasions or breeds that needed to be cut, or the nail gets too long, curls, and can cause a bird's health problems.

If you don't feel comfortable doing this, you can ask your local vet or live poultry supplier to help out.

If you're okay with that, you'll need a pet nail clipper.

Clip the nail, taking care as not to clip it too long or clip the pinkish part of the nail (quick) that will cause it to bleed and injure the bird.

Clipping wings

If you're worried about your birds flying in the coop and taking in your neighbor's yard, there's a chance to clip their wings.

This can be done by yourself (again, if you know how to do it and are comfortable with it).

It is best to first figure out how to do this or do it under observation, as you do not want to injure or permanently damage the wings of the bird.

Normally, just one wing clip allows the bird to become erratic when it tries to take off.

Each time the bird molts, you'll have to trim and slice the wing again as the feathers grow back

Dust baths

Chicken owners may find dirt patches that look like a chicken made a bit of a nest, but actually, it's more likely to be a dirt bath patch. The chicken is going to scratch a hole to loosen the dirt. They'll nest in and wiggle around, flapping out their wings and spewing up the dust. This is how they clean themselves up in a dust bath.

The dust bath helps to remove internal bacteria and extra oil from their feathers and hair, which helps to make them waterproof.

Dust bathing for chicken is just as normal human bathing is a way to groom and clean up.

The first time you encounter this, you might think that your chicken is in trouble because it's quite a sight to behold.

There are many ways to make a dirt bath for your chicken, and we have a few ideas for you in our "Chicken Dirt Baths" article. The article explains the forms of sand, nutrients, and the right dust baths for winter, etc. It is important to make sure they have some good loose sand to fluff around in the water, and you can also apply some peat moss to the heap.

Clean, warm, and comfortable in the cold and clean, hot, and new in the sun. The mean heart rate for chickens is usually 280 to 315 beats per minute.

Their average body temperature under normal conditions should be between 102 and 103 degrees Fahrenheit.

A hen will take about 30 to 35 breaths per minute, while a rooster will take about 18 to 20 breaths per minute on average.

The average lifespan for chickens depends on their laying habits and can range from 2 years to 20 years (this also depends on the chicken breed).

Because their bodies work differently from humans and other species, they seem to need some extra care at both the warm and colder periods of the year.

The hot summer months

OVERVIEW Now that you've decided whether keeping chickens is right for you, why, where, and how many chickens you can keep, the breed you want and age. It's time to look at what it takes to take care of your family.

Your birds should need quite a bit of your daily care and well-being.

TAKING CARE OF YOUR FLOCK: Your birds can take quite a bit of your daily care and well-being.

Keeping your bird(s) safe Vaccinations: When buying your ducks, it is best to tell the manufacturer what the birds have been vaccinated for.

If your chickens are combined with other poultry and domestic animals, it is best to have them vaccinated for Marek's disease.

It is also best to confirm vaccines for your bird(s) with your nearest vet, animal shelter, and or animal welfare society.

Prevention is always better than cure, so if you can stop an infection, it's best to do so, particularly if it's widespread in your community.

Chickens are also likely not to reveal that they're ill until it's too late. We must behave as normal as possible to avoid being harassed by the rest of the flock.

You can handle most of the vaccines yourself, but it's best to let a vet or an animal husbandry specialist do it.

Make sure you keep the vaccines up-to-date.

De-worming all animals kept indoors, or in a ranch, it should be dewormed at regular intervals, especially as these pests can spread at an alarmingly rapid rate to kill your flock.

Preventing / Treating worms: There are different methods to do this, but it's always best to check with your doctor for the right one for your flock.

We will also instruct you on how often to prescribe the drug to them.

Over counter solutions: One of the simplest and most cost-effective ways is running to about $20, it's a fluid wormer.

You put the amount you need in their drinking water about once a month.

Natural remedies: There are also more natural remedies that you can do throughout the month that will also help to get rid of any potential worms and even some other insects like sand mites and ticks.

You can try grinding the garlic with your meat and offering the pumpkin seeds, and even Apple Cider vinegar is going to do the trick.

Serious cases: In severe cases, it is always best to take the dog to the vet to ensure that you get the proper treatment to treat the rest of your family.

An epidemic of worms: if your flocks have worms, it is best to ensure that any other animals around them are still dewormed.

Make sure the soil on which they run, eat on, etc. is also worm-free.

Surroundings: there are materials that can sanitize the soil and ensure that the worms are destroyed.

Holding the grass short will help the sun's rays consume any remaining insect eggs that may still be in the grass.

There are drugs such as Soubenol, Flubenvet, and other herbal remedies that can be purchased from your vet and pharmaceutical distributors.

Symptoms of worms: It's not easy to tell if your bird(s) has worms, but in the early stages, they can leave to lay eggs in unusual places.

They may have diarrhea, with some worms, you will be able to see in the chicken's falls, they will feed more, they will lose a lot of weight, and if they have gapeworm, they may extend their necks as if they were gasping for air. Severe infestations can be deadly to the bird(s).

Forms of worms: Tapeworm is not that widespread but can happen and will damage the bird(s) immune system when they strike.

Gapeworm Photo Source: extension.umaine.edu. These disgusting worms are bound to the trachea bird(s), and that's why they make the bird stick out their necks and appear like they're gasping for air. Such worms are more common in chickens as they can be picked up from hosts, including snails, slugs, and worms.

Roundworm: There are a few varieties of these worms, the most common of which is the big roundworm. These can basically take place anywhere in the digestive system of the bird(s). You may be able to pick them up in your bird's drops.

There are other types of insects other than worms that can damage your flock.

These are quite normal and may happen even with the regular care of your coop/run. Fleas enjoy the warm weather and can be quite a nuisance in the summertime, particularly in the hottest climates.

Prevention / Care of fleas, mites, and lice: washing and care of coop, yard, garden areas, and other domestic animals that may come into contact periodically is a must.

Check your birds often by inspecting their feathers close to their skin, particularly on their breasts, tails, and vents.

Several mites are going to get between the scaly areas of the chicken legs.

If they have one of these infections, they can be handled with different solutions depending on the parasite.

Over the counter solutions: you can get any supermarket or pet hospital over the counter medications.

Please read and follow the guidelines, if you are not certain that you will seek advice from your nearest vet or animal shelter.

Ensure that the areas in and around the coop or wherever the chicken's play has been cleaned.

Natural remedies: There are garlic sprays available, or a person can make their own by smashing 3x whole cloves of garlic in 2 cups of water. You can also add a teaspoon of lavender, spearmint, thyme, berry, or cinnamon oil to the mixture.

Spray chickens, coop, farm, and a number of locations around the coop.

You can even apply crushed garlic to their water or dry food.

Extreme cases: In severe cases, it is always best to take certain types of flea from the animal to stay near the eyes of the bird, and a serious infestation of these can lead to blindness. If the chicken is in severe discomfort, it is safe to have the infestation handled by a vet.

An insect outbreak: When one bird has fleas, it is almost certain that the rest of your flock has fleas. Every bird must be handled, its coop fully washed, just like the surrounding areas.

Ensure that all the animals are handled as well.

Once you've washed the coop, make sure you take off any socks, boots, etc. before you get back. These should be washed and cleaned up afterward, as you don't want to catch them in your house.

The surroundings: make sure you keep the grass low as these insects appear to grow in the lawn.

There are both synthetic and organic methods in the greenhouse to get rid of the insects.

Symptoms of insects such as mites, fleas, and lice: All three species tend to show the same symptoms of infestation. Your birds should exhibit the following most common symptoms that may be aggravated by these parasites, pain and blood loss, Decrease egg production, Constant scratching, Irritability, General malaise, Restlessness, See feathers deteriorate in patches and have damaged skin spots. Many infestations can cause blindness and even death. The insects that you may find with your bird(s) are Mites.

They do come in three varieties:

Northern Fowl Mite: they love to live in the Coop's vent area, but they are quite happy to live on chicken as well.

Red Mite: these are the most annoying mites, and once an infestation happens, it's really hard to get rid of them. You may even need to get a qualified pest control to get rid of them. We love to live in nest boxes and other coop nooks and crannies. They're going to feed on chicken's blood during the day.

Scaly-leg Mite: these horrid little critters live in the scaly areas of the chicken leg. This can make the bird go lame if it's not handled on time. They're going to make the chicken leg thick, rough and swollen.

Lice: These nasty critters feed on chicken feathers and hair flakes. These are very irritating to the hen because these lay their eggs at the base of the feathers. Many situations are so serious that the bird is unable to collect the eggs.

There are some 50 different kinds of lice that can be present on poultry. Ask your local poultry specialist for more in-depth information on these.

Fleas Most of the chicken fleas you can see are only about 5 percent of what is currently on the bird. The remaining 95% or the flea species is either larvae or egg-shaped on the animal.

Fleas can live in a long grass for a short time, but they need a host to survive. They live on a bird and tend to breed rather quickly as soon as they are well fed.

Consider sprinkling FGDE on the coop wall, nesting boxes, roosting areas, and it can also be dusted on your flock (always check the directions before applying it).

While garlic is excellent as a general deterrent to most pests and worms, there is a risk of Heinz Anemia. This is because garlic is part of the family of onions. Onion is not a good source of food for your pets.

Apple Cider Vinegar sprinkled gently on your flock feathers as well as some herbs like mint will keep them away.

Ticks Poultry ticks are sly critters that you may not spot on the bird. That's because they're hidden in the coop during the day to feed their host at night.

Such ticks are, in addition, borne by wild birds and are very common in nests such as vultures, buzzards, egrets, etc.

They live in dark crevices between the wooden slats of the coop or in cracks in the floor.

They love the heat, the damp, and the humid air. Although the really cold weather is going to kill them, it's best to keep your coop tick-free no matter the season.

The colder winter months

Most chicken breeds survive the winter without any complications, as their feather allows for great insulation. There are also some types of chickens that do not take too well in extremely cold climates, so it is best to check the breed and assess their temperature toughness.

Nevertheless, the cold can have a serious effect on even the toughest animal, particularly when they lie down for the night.

Look out for the following signs that your chickens are suffering from the cold weather.

Egg production decreases: The birds will huddle together in groups. There will be a rise in their appetites. They will not consume as much food. Your birds may have behavioral changes. They will not be able to ward off infection as well in the snow.

They may have problems with their feet if they are constantly walking around in the damp. Older birds will be more affected by the cold. Some birds may suffer from frostbite. Help prevent damage to your flock and keep them warm by adding energy-rich fats and oils to their diet (see your local vet for advice). This will also reduce the consumption of food for birds in the cold. Provides a good source of energy to help the bird keep warm. Make sure there is no excess of groundwater, as the birds may be tempted to drink, and that could lead to worms. It is particularly good to ensure a rigid de-worming routine in these cold months. Something that you need to discuss with your local vet.

Isolate their coop and keep it elevated from the ground, keeping the birds safe in the rain.

Heating lamps are a good option, but they also come with their dangers. It's always best to check these as they could be a fire hazard.

Keep a close eye on your birds for any winter malaise, so that you can deal with the situation early and avoid any future outbreaks or dire consequences for your bird(s).

Keep your flock clean, sanitized, and keep healthy.

Feeding meals and foraging feasts- There is a clear rule of thumb for feeding and handling the flock, and this is the 90/10 rule. This is to ensure that 90 percent of your chicken's average intake of food is their healthy feed, with 10 percent left for delicious treatments.

We have a minimum of 38 different nutrients at the right level. These are typically found in different "Full Feeds" for poultry.

Feed your chickens in the morning before they go to forage (every feed pack must come with instructions on how much to feed your chickens). This way, you're going to make sure you stick to the 90/10 percent rule. Chickens will have their normal nutrient requirements in their feed before they go out to peck and forage. Foraging foods are also considered to be part of their therapies.

Full feeds come with all the bird nutrients you'll need. A common mistake that people make is trying to replace these feeds.

It's better to give some form of the substitute as a medication instead, along with the medications, just make sure you're not over-supplementing your bird(s). It is best to talk to a vet, animal shelter, or a poultry farmer for any advice.

Feeding the chicks: chickens from day 1 to 18 weeks old Use a complete starter feed.

Do not give treatment until the first egg has been laid or the roosters are older than 18 weeks.

Laying hens: Laying hens need a total of 38 nutrients in their complete feed, of which they should have at least 0.25 (half cup) pounds of each day.

You can get a variety of Layer feeds for birds producing table eggs.

Broiler feeds for birds that contain hatching eggs.

Feed the chickens in the night. Use Complete feeds for adults and chicks as described above.

Scratch grains are more likely to be treated and should NOT be mixed with their regular food.

Ensure that you provide enough grit to the chicken to help digest any grain or food that might require it. You can get the stuffed grit that's chicken size, fine gravel, or rough soil.

Oyster shells are too small to make a good meal, and you need to be careful when feeding them. Chickens do not need too much calcium and may have an adverse effect if they overdose.

Foraging

Chickens love to forage and try new plants and little bits they can dig around the garden.

They're quite in tune with their brains, so they don't tend to eat what isn't healthy for them. But there are some chickens that are going to eat just about anything, so you can take care of the kind of flowers, herbs, berries, etc. that you may have in the greenhouse if you keep chickens.

You can throw a tablespoon or two of the scratch grains around where they feed and scratch during the day. It's best to do that in the afternoons.

What chickens can eat: The darker green leafy vegetables tend to produce a darker egg yolk.

A few healthy favorites include; perennial favorites, roses, ferns, coneflowers, daisies, hostesses, and daylilies.

Vegetable favorites; kale, Swiss chard, carrots, lettuce, beets, and broccoli.

Fruits Favorites: fruit should be seasoned Strawberries, blueberries, avocado meat, and watermelon Favorites: cilantro, thyme, basil, mint, oregano, parsley, and lavender. Then there are foods that contain toxins and substances that can make them sick or even cause death.

Avocado pits and skin contains a poison called persin, which is poisonous to chickens. No moldy or rotting feed and food containing a lot of salt, as this may be harmful to your flock.

No meat products Rhubarb with severe weather exposure typically contain a high concentration of oxalic acid that is poisonous to poultry. Rhubarb also produces anthraquinones that have a laxative effect on your birds.

Try to avoid beans of any kind, particularly undercooked ones, as they may contain something called hemagglutinin, which acts as a digestion inhibitor.

Onions are not only harmful to your birds, but they can give their eggs a strange taste as well.

Garlic may be provided in abundance to protect against certain insects, but keep in mind that it may have an effect on the flavor of the eggs.

Chicken friendly garden. If your birds are going to be free in your garden, it's best to take care that it's a chicken-friendly garden.

Keep in mind that it's like scraping the ground, foraging, and eating plants.

Any plants or areas you don't want them scratching inside, you might want to look at putting chicken proofing in the form of tunnels or chicken wires around them. Especially if you've got a vegetable patch or a herb patch.

Keep the crops as chicken safe as you can.

A good old chicken time and general maintenance.

Spending time with your flock every day: Chickens are really clever and quite sociable animals. Most of the breeds will flee when they see their humans entering the yard. We also like to lead you around thinking about their day.

As with any pet we want love, you can spend at least an hour or two with your herd.

Even if you're just walking through your garden with them trailing behind you, give them a word, or if they want a cuddle. Inspect their feathers for the critters to make sure they look after themselves.

They like to be chewed on the back of their necks—gently, though!

You could even play with them and show them tricks.

It's very important to develop a trusting relationship with your bird(s).

General cleaning: Every day, feed the birds each morning before they let them fly, giving them treats in the afternoon and early evening.

Save eggs (collecting them daily will prevent them from being eaten). Fresh food (make sure there are no falls or foreign objects in the bowl)

Fresh water (make sure you don't kill them).

It's a great idea to add some beautiful herbs to their bedding, such as lavender, lemon balm, and some fresh mint oil.

Give the birds a thorough check. To do this, gently pick up the hen and start working your way down to their feet with your head.

You need to test for weight loss, sores, fleas, mites, watery eyes, or the odd color of your combs. There should be no scab or hard-swollen vine. The hen's wind is usually moist and pink. The legs should be smooth, not scaly, and there should be no black spots. Their breast bones should not stand out that could suggest weight loss.

Once a month

Do a good thing to wash the coop.

Replace all the bedding and wash the nesting boxes thoroughly.

Wash all of the things in the coop, such as roosts, feeders, and water bowls.

Scrap any drop and completely clear out any stray, line the floors, etc.

Check ventilation and vents to ensure that the coop is adequately cleaned and correctly sealed.

Replace and replace any chicken wire that may be required, as well as various items such as nesting boxes, etc. in the coop.

Each six months

De-worm chickens Test vaccines Ensure that special summer and winter precautions are taken

Once a year

Remove all heavy equipment such as lights, fans, coops themselves, etc. if necessary.

If you can, have a vet check up on your flock to make sure they're healthy.

CONCLUSION

Raising a chicken flock at home can be a good experience and a source of fun. This talks about living beings and accountability as a family project. Home chicken flocks can also be an excellent source of low-cost, high-quality poultry products. This publication should provide the basic tools for starting a successful flock.

Andrew McDeere

Raising Goats

A complete Guide to Raising, Breeding, Keeping and Taking Care of your Goats

Copyright © 2020 publishing.

All rights reserved.

Author: Andrew McDeere

No part of this publication may be reproduced, distributed or transmitted in any form or by any means, including photocopying recording or other electronic or mechanical methods or by any information storage and retrieval system without the prior written permission of the publisher, except in the case of brief quotation embodies in critical reviews and certain other non-commercial uses permitted by copyright law.

Raising Goats

Raising goats can be costly and monotonous, however, in the event that you know the nuts and bolts of goat care, it will make life a lot simpler for you and your creatures! In the event that you love your goats, you need to be sound and glad. There are approaches to accomplish these objectives.

There are several types of goats. Prior to getting, you may solicit yourself: "What sort of goats would I like to lift?" For instance, there are dairy goats, meat goats, pet goats, and pet goats. Snow-capped goats and Saanen are instances of some huge milk makers. Boers can be reared for meat or show. Small scale goats or Nigerian Dwarf goats can be reproduced for show or as a pet. Blacked out goats are fun livestock! Raising these creatures can be troublesome and costly, so there are numerous realities to consider.

Take a gander at the earth and things around you. Is your condition helpful for goat reproducing? These animals like to walk and touch in enormous fields, as can be found in numerous homesteads. They ought to never be tight or contained in little regions. Does your condition have sufficient structures around you? Without a doubt you would not need your goats to meander, get lost or torture your neighbors. Fitting structures could incorporate various sorts of wall and Gates.

Goats need a safe house, a little outbuilding or a shed. In spite of the fact that they function admirably in chilly conditions, they can be delicate to delayed warmth. Your haven ought to be clear, perfect and liberated from drafts. The floors of your safe house ought to stay dry as goats might be inclined to foot decay. Keeping up a decent and clean sanctuary will decrease the spread of microscopic organisms. This will help keep a solid creature.

Taking care of your goats with the correct food and setting it accurately is significant. All things considered, what is the purpose of having the best nourishment for your goats, which you can discover, however it was not reasonable for them to devour? A food blend is utilized to take care of numerous goats, yet the healthful needs of your goats may should be chosen by an expert, for example, an approved veterinarian. An expert or neighborhood raiser can assist you with learning the perfect measure of food to give your goats and the opportune chance to take care of them. Water ought to consistently be close and available to these creatures.

Keep up the strength of your goats with preventive consideration. Know your creatures and their every day schedule. Note if there is an adjustment in your appearance, conduct, or disposition. There are numerous signs that could mean an issue. For instance, a murky layer may show bugs, decline to eat, don't focus, or having a fallen tail could mean issues. These are only a couple of guides to stay informed concerning the state of goats, there is significantly more.

Ensure your goats remain sound. Track your immunizations and clinical history. Make a nearby and careful check of your jacket, eyes, feet, and sack. Make a general and customary check of your appearance, disposition, and conduct. Keeping up appropriate consideration of the goat will permit them to carry on with a long and glad life.

Make certain to invest energy with your goats. Caress. Talk in a quiet voice. Goats don't care to be separated from everyone else constantly. They appreciate the organization of different goats and individuals as well! Raising a goat probably won't be a smart thought. Having an organization will assist them with remaining cheerful.

Raising goats merits the work and exertion. They can be amusing, fascinating and entertaining creatures. Whatever sort of goats you get up, knowing and keeping up fundamental goat care will make your life simpler and assist you with keeping your goats sound and upbeat.

RAISING GOATS PAST AND PRESENT

The history of goat farming goes back nearly 10,000 years in Africa and the Middle East. Its use has remained relatively equal over the centuries, people around the world raise goats for their meat, milk, hair and usefulness as load animals due to their agility and confident feet.

Goat meat and milk are consumed almost all, this is a daily diet, especially in the countries of the Middle East. Milk is processed into cheese and other food products, skins used as a material for clothing, housing and containment of liquids such as water or wine.

Goats are also great pets, which the first goat keepers learned quickly as they spend much of each day with their flock. The shepherd brought his goats daily to an area that provided plenty of fresh grass for grazing and clean water, keeping them from any predatory animal that could hide. Every night, the shepherd gathers his goats in the barn and locks them for safety.

Modern times have left this ritual of raising goats relatively unchanged, for the most part. Fences and automatic grazing and irrigation systems for those who have the means replace the daily tasks of the GOAT, but many tasks still need to be performed manually, such as administering medicines against the disease and maintaining the correct nutritional foods available.

In the past, there were only a few different breeds of goats in the world. Today there are many different breeds of goats that cross and breed thoroughly. Although there are many breeds of goats, only a handful are popular for several reasons. Breeds include Boer, Alpine, Toggenburg, pygmy, Spanish, Nubia, fainted goat, LaMancha, Angora, Kashmir and more recently, the Kiko goat which comes from New Zealand.

Although there are several reasons for raising goats, the possibility of pleasure and profit remains the same everywhere, whether the goats are for the companion, dairy products, meat or fiber.

RAISING GOATS TERMINOLOGY

Goat terminology used to denote the different characteristics related to various aspects of goat farming. There is a word for everything from sex to age and appearance. Learn the words below and their definitions to understand what other goat owners in the world are talking about!

Some terminology is the same as deer, since goats are related to deer, if you know something about deer, you may recognize some of the words below.

Doe or nanny is a female goat.

A dollar or a billy refers to a man.

When a male has been castrated or castrated, it is called a wether.

A young goat under one year old is called a child.

A group of goats is called herd.

Chins are like small bags that sometimes hang under a goat's chin around its neck.

Goats are commonly used to refer to goats in general, and is derived from the scientific name goat aegagrushircus.

Musk is the strong smell emanating from males during the furrow, which is the breeding season.

The goat brush is a cross breed, this word is the equivalent of "mutt" in the world of dogs.

The hair used in the production of fabrics and yarns is called Cashmere and also refers to the breed commonly used to obtain it.

Goat meat is usually called chevon, which comes from the French word GOAT; if it is the meat of a child, it is said to use the Spanish word GOAT.

This is not the extension of goat terminology, but these are some of the most commonly used in the industry. We hope that this information has helped you get acquainted with goat breeding, and you will soon become an experienced owner.

Your Guide To Goat Farming

Goats produce two very important products in goat farming: milk and meat. In most large goat farms, goats are treated as dairy cows, as they are housed indoors and milked twice a day. Large farmers have more than 400-500 goats on their farms.

The breeding season of goats on farms runs from August to March. The pregnancy of the goat lasts four months, and they are usually bred once a year, so their children are born between January and August. Female goats give birth to one to five children and twins are provided.

A female goat on a farm can begin mating after the age of seven or nine months, while it can be milked when the goat reaches a year. Goats give birth easily, so no special help is needed. However, farmers should make sure that children breastfeed from their mother, if they do not, they should be fed with a bottle. This should be done immediately after the birth of the baby, as it is when he receives the first critical milk called colostrum. After feeding colostrum containing minerals, vitamins and antibodies for a few days, the baby could be fed artificial milk or could breastfeed his mother.

Breeding goats on a farm is quite similar to breeding cows. Goat babies should receive a milk formula until they can be weaned; this is after they reach five to seven weeks of age. This is the time when goats are milked.

On a goat farm, females are given a period of two months before delivery, they need this time so that they can feed their children after birth. To milk goats on daily farms, goats are milked twice a day, usually at intervals of 12 hours. Milk can be extracted by machine or by hand depending on the type of techniques and the workforce of the goat farm. Another thing that makes goats and breeding cows similar is that they both use up-to-date magazine production that must meet certain hygienic requirements.

If the farmer is more interested in meat production, goat babies should be breastfed for eight to ten weeks. After that, they should be fed hay, cereals and pastures until they gain enough weight, which can range from 35 to 90 pounds.

When a farmer breeds goats for his meat, he must consider the breed of goats, and then decide what optimal weight the goats should reach. Different breeds of goats reach a different weight. Goat breeding may not be the first thing to consider when talking about agriculture, but it is a profitable and pleasant business.

Raising Goats As Pets

Albeit regularly portrayed as a ranch or working animals, a little goat homestead can be extremely remunerating for the proprietor and is generally simple as long as you keep some essential principles. A little goat ranch can be a superb wellspring of milk and meat for the proprietor, and raising the goats yourself, you can be certain that they were brought up in a sound manner. Goats can likewise keep their territory for all intents and purposes without grass.

Goats Are Social livestock and you should plan to permit at any rate two goats to live respectively. The best varieties to keep differ contingent upon whether the fundamental motivation to keep them is milk, meat or fiber, or whether you need them principally as pets.

Realities of the goat

Goats are known as guys and females do. Baby goats are called Children. Goats normally live from 10 to 12 years, in spite of the fact that there have been instances of goats satisfying 15 years. There are in excess of 300 separate types of goats and are all the more firmly identified with sheep, with which they can cross the variety, despite the fact that this isn't suggested. The principle items related with goats are milk, cheddar, meat, mohair and cashmere.

Goat items

Goat milk is turning out to be progressively well known and an enormous dairy goat can deliver 3,000 to 5,000 pounds of milk every year (2 to 3 liters day by day). In many areas, milk should be sanitized in the event that you need to sell it industrially, despite the fact that it is conceivable to drink untreated milk from your own goats. You ought to know that there is research recommending wellbeing dangers with the utilization of unpasteurized goat's milk. Likewise with milk, there is a developing interest for goat meat and it is asserted that they are medical advantages contrasted with other red meats and chicken. On the off chance that you sell meat, you should adhere to the standards that a little business processor must follow. The standards are less severe if the meat is planned for utilization. Some goat proprietors think that its more helpful to redistribute the butcher to an approved Butcher. Goats were additionally refreshing for three sorts of strands, mohair, cashmere and cashgora.

Goat convenience

It recommends a free and dry development venture that shields them from the components and gives adequate security against rodents and different predators. Rodents could present the infection, just as eat and tartar food and water holds. Concerning size, there ought to be sufficient space to permit goats to remain on their rear legs with their necks resting. Independently, every goat ought to be around 4 square meters. M.M. territory. On the off chance that the goats are housed in a gathering in a similar region at least 2 sq.M. M. M. for the GOAT, it ought to be given, albeit more than this base is prescribed to keep away from clashes. Goats with horns and with unsettles or without horns ought to be composed independently.

Goat food

Regardless of whether they have gained notoriety for eating nearly everything, they won't flourish except if they have the correct adjusts in their eating routine. Regardless of whether they eat weeds and different plants, including pastures, they will require access to great quality roughage. Vegetables contain more minerals, nutrients and supplements, in spite of the fact that, similarly as with different roughages, quality may fluctuate contingent upon assortment, planning and capacity.

The GOAT of Health

There are various maladies that can influence a goat in both ceaseless and treatable structure. A portion of these illnesses can be sent to people and different animals, while a few ailments are explicit to goats. Two ailments that can prompt unexpected demise in a goat are coccidiosis and pneumonia. A large portion of the worries for reproducers and makers are worms and nuisances. A goat that covers with bugs and worms and isn't dealt with will probably endure a quick decrease in wellbeing, creation and frequently cause demise.

Various Types of Goats

Goats are currently tied in a wellbeing cognizant society, on the grounds that their milk contains better fats and proteins, just as their meat has a high healthy benefit. There are a few sorts of goats that people breed for ranch, meat, milk, fleece and different purposes. Dairy goats, Nigerian dwarfs, Boers, Kashmir, dwarf goats and mountain goats are a portion of the fundamental sorts of goats you will discover in different pieces of the world.

Smaller than normal goats are little goats reproduced from breeds, for example, Cashmere, Australian goats, Angora, Nubia, West Africa, and so on scaled down goats are delicate, neighborly and delightful. They are interested, terrible and savvy. The fundamental preferred position of smaller than normal goats is that they require less space.

They like to invest more energy in outside conditions, and thusly it is appropriate for appearance in the patios of houses. Scaled down goats have a decent future as long as twenty years relying upon the given consideration.

There are small dairy goats with Nubian ancestries, Saanens and Toggenburgs, and so on smaller than usual dairy goats are reasonable for little homesteads and houses that don't have huge zones. One preferred position is that they need less food against enormous goats, yet they could give great volumes of milk. Smaller than normal goats impact animals, for example, ponies, cows, and so forth. In this way, they are generally used to go with these animals.

These goats get exhausted without any problem. Consequently, we invest quality energy with them. Appropriate sustenance and normal assessments with veterinarians will help keep up the great soundness of these goats. You can purchase miniatures of goats at a youthful age from four to twelve weeks. You can jug and you get the feeling that they are more similar to canines. Smaller than normal goats like to have a decent brushing meeting.

Dwarf goat is the most famous and entrenched smaller than expected goat type of West African source. They look charming in their short covers. They are spry, littler and full. They produce a volume of milk in spite of their little size. In any case, they are not favored for developing trunks for milk and meat purposes. Dwarf goats duplicate during the time ceaselessly.

They weigh around twenty-three to thirty-four kilograms. State that guys weigh from twenty-seven to thirty-nine kilograms. Dwarf goats are there in various hues, for example, strong dark, different shades of caramel and shades of agouti.

Dwarf goats need cleaner day to day environments. Your eating routine incorporates grain, oats and clean water. The eating routine for every goat will rely upon its age, size, and so on you can oblige dwarf goats in a little shed or a bigger specialty. Give fitting sheet material the flavors of sawdust, straw or elastic mats. Male dwarf goats can get forceful as they become grown-ups, so disinfection is important.

You need a normal cut of the head protector and an exit from the body. They are alright with low temperatures up to twenty degrees. Overseeing dwarf goats is simpler, also, they are delicate, lively and loving. These goats are principally appropriate as pets and useful for presentations and fairs.

KEEPING GOATS

The interest for goat items means that great possibilities in goat cultivating. There are some significant realities to remember when you begin developing goats. You can settle on dairy goat cultivating, which centers solely around milk creation or meat goat cultivating which manages meat creation. Choose what number of goats you need to have on the homestead, contrasted with the complete space on the ranch.

Goat lodging is a significant part of goat cultivating. Area with low and precipitous zones that are a long way from the expressway and appreciates great water system, quality air, shade of trees, and so forth goat lodging must have a decent tallness for goats to stand high. There must be acceptable ventilation, adequate space for food, an appropriate seepage framework.

Lodging ought to give assurance against wild animals, climate conditions. The territory of in any event four square meters is required for a solitary goat. There is likewise requirement for new water gracefully, draining space, dry space to continue taking care of, and so on you need a decent dry floor with legitimate sheet material.

At that point select the best type of goat by checking the heredity and breed, compliance or body shape. Screen development design, milk creation limit and fruitfulness before choosing goats for ranches. South African Boer, Nubians, Tennessee goat meat, Kiko, and so on are a portion of the great assortments of goat meat. Elevated, Nubia,Toggenburg, LaMancha, Sarine, and so forth are a portion of the types of goats for dairy cultivating.

It is essential for domesticated animals the board to deal with goats under different conditions. On account of doing', unique consideration is required during pregnancy until the joke. Male reproducers, youngsters need diverse consideration. The administration of swelling of goats is basic on account of meat creation.

There must be a suitable administration program for all the exercises of the organization. There ought to be adequate work force to perform such exercises as taking care of, cleaning, haircutting, deburring, foot trimming, haircutting, detachment, and so forth. Since the odds of disease are higher, a viable cleaning instrument is required.

The free brushing framework is reasonable for ranches, which are bigger, however the administration of road goats is a major errand. Letting the goats nibble during the days in the controllable space will be acceptable, as this will assist them with munching great grasses and grass. Youngsters will have the chance to run and feel the glow of the sun.

The goat diet ought to contain proteins, starches, nutrients, minerals, and so forth., the goat diet dependent on grains and vitality enhancements will be helpful for keeping goats solid. Uncommon food is required for goats proposed for meat since fat goats pull back more cash.

For domesticated animals ranches can utilize reproducer dollars or managed impregnation relying upon the accessible offices. On account of dairy ranches, machines for draining goats are required. Great milk storerooms are required.

Full-time veterinarians and standard wellbeing checks are required. The exchange of wiped out individuals will help forestall the spread of the infection to different goats. Keeping goats involves fun, since they are delightful animals.

Multiple Uses of Goat

Goats are extremely valuable both when they are alive and much after death, offering meat and milk as the skin offered by the skin.

A cause is engaged with giving goats to the poor in Africa.

The primary preferred position was that goats are anything but difficult to oversee contrasted with domesticated animals and have various employments.

MEAT

Goat meat is called GOAT, which is like sheep meat. In any case, some accept that it tastes like meat or deer, it relies just upon the age and state of the goat. It can likewise be set up in different manners with stew, heated, barbecued, flame broiled, slashed, canned or even arranged in hotdog. Goat jerky is another darling assortment. In

India, rice = readiness of Biryani utilizes goat meat to create a rich rice flavor. As far as sustenance, it is lower in fat and cholesterol. It conveys a bigger number of minerals and diminishes complete immersed fats than some other meat.

Different pieces of the goat, including the organs, are entirely palatable. Exceptional joys incorporate the mind and liver. The head and legs of the goat are smoked and used to get ready fiery dishes and restrictive soups.

MILK AND CHEESE

Goat's milk is all the more effectively processed by people and is suggested for the most part for youngsters and individuals who experience issues with bovine's milk. Solidifying arranged from goat's milk is a lot littler and more absorbable. It likewise homogenizes in nature, since it does not have the protein agglutinin.

Goat milk whenever took care of effectively, from perfect and solid goats, sound and cooled as quickly as time permits, the taste is unimportant and innocuous. Likewise, it is important to isolate the solid smell from the dairy, as its fragrance will rub them and stain the milk. Goat's milk is utilized to make extraordinary worth cheeses, for example, Rocamadour and feta; regardless, it could be utilized to make different sorts of cheddar.

FIBER

Cashmere goats produce the best fiber, cashmere fleece is truly outstanding on the planet. Cashmere fiber is incredibly slim and delicate and develops under gatekeeper hairs. The cashmere goat was especially high to make an a lot bigger sum with less Guard hair.

The Angora breed creates long wavy and sparkly mohair strands. Locks develop consistently and could be four inches or considerably more long. Goats ought not be slaughtered to cut the fleece that is cut on account of Angora goats, or to brush on account of cashmere goats.

In South Asia, cashmere is known as pashmina (Persian pashmina = fine fleece) and these goats are known as pashmina goats. Since these goats, truth be told, have a place with the upper locale of cashmere and Ladakh, their fleece was called Cashmere in the West. Cashmere wraps pashmina with its group weaving is extremely popular.

SKIN

Goat skin is utilized today to make gloves, boots and different items, which require delicate skin. Kids' gloves are trendy in the Victorian time are as yet made today. The Black type of Bengal, neighborhood in Bangladesh, offers a top notch skin. The calfskin is likewise utilized in Indonesia as a carpet and instrumental drum made of neighborhood cowhide called bedug.

Different pieces of the goat are likewise consistently valuable. For instance, the digestive tract is utilized to make catgut that remaining parts the favored material for inward human stitches. Goat horn implies prosperity (Horn of bounty) is likewise utilized for making spoons, and so on.

Guide to Raising Goats

No ifs, ands or buts, goat cultivating is one of the most gainful homesteads today. Notwithstanding selling new and handled meats in the business sectors, there is additionally goat's milk that can be gathered and sold new or utilized as elements for different nourishments (for example desserts, cheddar, yogurt, and so on.) and healthy skin items (for example salves, cleansers and creams. The strands of these animals additionally produce fleece, mohair and cashmere fleece; and now there are ranches that breed and sell easygoing goats as pets. In the event that you are thinking about raising goats as a business endeavor, here is Guide 5 to raising goat tips you should seriously think about.

Goat rearing aide Tip # 1: Consider what kind of creation you need to enter. Would you like to sell goat meat, milk, fiber or pets? Obviously, you can sell milk and goat meat simultaneously (or any blend you need.) But this would include enormous by and large consumption from the beginning. It would likewise mean getting heaps of animals and a similarly huge ranch space. Attempt to begin this independent venture. This will help minimize your costs while learning the ropes of business scale goat cultivating.

Goat reproducing guide Tip # 2: Now that you have picked the kind of creation you need to enter, you ought to painstakingly pick which variety of goats you can arrange. Goat breeds, for example, Angora, cashmere, Nigora and Pygora are brilliant for fiber creation. The best goat meat makers are: South African Boer, Kiko, brush, myotonic (otherwise called swooned goats), West African dwarf and Spanish goats. Goat breeds, for example, Alpina, Anglo-Nubia, La Mancha, Saanen, Toggenburg and Oberhasli are the best milk makers; while submissive varieties, for example, Anglo-Nubia, South African Boer and dwarf goats can be reared and sold as pets.

Goat cultivating guide tip # 3: gain proficiency with everything you can about business goat cultivating. Pursue goat cultivating courses and how to gather and sell goat items. Approach neighborhood goat ranchers for counsel and business exhortation. The more you think about this kind of business, the quicker you can recoup overhead and advantage from your difficult work.

Goat cultivating guidance # 4: consistently look for the administrations of a veterinarian. Generally speaking, goats are solid animals and furthermore have a genuinely low support. Yet, in the event that you sell meat and milk from animals, you need the animals to get spotless wellbeing bills. Furthermore, having a veterinarian on your ranch installment is compulsory in many states.

Goat cultivating guide tip # 5: form appropriate lodging for your animals. Goats would require assurance against likely predators and components. To flourish, a goat would require in any event 4 meters of indoor space with a lot of free space to stand up to. Lodging ought to likewise incorporate a different taking care of territory, watering station, trash and draining or cutting stations, particularly on the off chance that you raise goats for milk or fiber creation, individually.

THING YOU NEED TO KNOW CONSIDERATIONS BEFORE YOU BUY YOUR OWN HERD OF GOATS

The capability of goat cultivating as a business industry has not been completely abused because of the basic actuality that individuals despite everything incline toward dairy animals' milk to goat's milk. Be that as it may, ranchers understood that goats require less support than bovines. Therefore, there was a steady increment in the quantity of outbuildings that started to develop and raise goats. Goat reproducing, both for pets and for business, started to be known as one of the most helpful choices for rearing.

Be that as it may, before you have the goats on your property, you have to ensure you have the assets to give them all the requirements they need. Here is a snappy guide that you have to realize that comprises of a few focuses that you ought to consider for rearing goats. Ensure you have thought about everything before purchasing your own clump of goats.

Goat reproducing guide Tip 1

To begin with, you have to have enough land for your goats to live, play and feed. Two sections of land of land are suggested for every goat. Man sure to pay every goat with its own space and abstain from packing all in a little territory.

Manual for goat rearing Tip 2

At that point you have to manufacture a safe house for domesticated animals. It's not really an extravagant stable. Everything with dividers and roof will be sufficient. Make certain to be outfitted with stages for dozing, where your goats can rest, and feeders, where they can eat and drink. Keep your sanctuary clean consistently. This will keep your goats from contracting undesirable maladies.

Goat reproducing guide Tip 3

Do your exploration on the dietary needs of the variety you mean to increment. Ensure you approach goat nourishments and enhancements. They as a rule require a consistent utilization of feed and water. To ensure they develop solid, you can likewise give them horse feed supplements routinely.

Goat reproducing guide tip 4

Lastly, he constructs a fence around his brushing region. You sure would prefer not to walk, isn't that right? This will keep their goats on their property and get predators far from their domesticated animals. A decent fence will give your goats all the insurance they need.

3 ESSENTIAL FACTS TO RAISING GOATS SUCCESSFULLY

Goats have consistently existed since antiquated occasions. They are regularly raised on ranches as a wellspring of milk, fiber, spread and meat. Their skins and hair were additionally gathered and sold in the business sectors. The male goat is frequently called "buck", while the female goat is designated "doe". As of recently, they are as yet delegated pets, however a few people embrace them as pets; outlandish animals really, in light of the fact that they are generally livestock and are kept in crowds. In any case, because of their normally inquisitive and friendly being, a large number of them end up as such nowadays and how to raise goats leaves them an undertaking instead of filling in as when rearing wild animals and reptiles.

Actually, there are individuals who are keen on reproducing goats as pets, while a few ranchers are thinking about making a goat ranch on account of benefit. Whatever the aim, every proprietor ought to comprehend that goats need appropriate consideration and care. Here are the fundamental qualities of goats, which are their signs on the most proficient method to raise goats.

Nourishment

Raising goats is anything but an extremely relentless employment since goats are not touchy to food. They can bite anything they find, however be mindful so as not to take care of them with wilted natural product, as it very well may be poisonous to them. On the off chance that a belladonna plant likewise develops on it, goats ought to be expelled. Fundamentally, it is smarter to plan spices and clean water to take care of them. Now and again they bite a can without gulping it just to check on the off chance that it's consumable, and afterward drop it a short time later.

Atmosphere

They are as yet delicate to outrageous climate conditions. Leaving them cold and sticky for significant stretches puts them in danger for pneumonia that regularly prompts demise. Quick treatment is fundamental at the main indications of the malady to draw out its term. Typically, goats have 15-18 years of future, yet there are situations when this expansion with appropriate sustenance. Therefore, as a matter of first importance, it would help you a great deal in the event that you are still at the phase of considering how to raise goats.

Friendliness and conduct

Goats develop inside a crowd, which implies it's an Animal gathering. So confining them would cause them Depression. This condition could likewise be lethal. It would not be an issue for ranchers, as they expect to raise a group. Pet proprietors can have in any event two to give themselves a friend.

Raising goats is a fun and productive business and it doesn't take any muddled recipe to recoup sound dollars and that is the situation. Continuously recollect these fundamental realities about them and you will prevail in your exploration on the best way to raise goats.

Why Should You Raise Goats?

Goats are probably the most established creature that people have kept and trained. They showed up and were portrayed in history books and even in the Bible. Today, their utilizations are still fundamentally the same as those where individuals restrained them. In the event that you live in the open country, it is prescribed to think about rearing goats.

Reasons why you should raise goats

On the off chance that you visit a ranch, the standard animals that invite you are chickens, chickens, bovines and ponies. These animals have incredible monetary and residential significance. In any case, on the off chance that you need to present another variety of animals from your homestead, you can attempt to put resources into goats.

Why raise goats? These are the primary reasons.

1. Goat meat gives you enough food.

In light of their little size, goats are a more favored wellspring of meat than other bigger animals, for example, dairy animals and pigs. Truth be told, these animals don't occupy an excessive amount of room, so they are simpler to store or store. Also, while enormous bodies compel you to spend more hours thinking and executing protection strategies, you barely need to do this with goat meat. Your family may have adequate supplies of goat meat until the following butcher.

2. It can change over goat's milk into various sorts of dairy items.

Goat milk has much preferred quality over milk delivered by different animals. Specialists state that it is more nutritious and is perfect for nursing youngsters who are susceptible to dairy animals' milk. Furthermore, in the event that you realize how to transform goat's milk into other dairy items like cheddar, you can get a ton of advantages.

3. Goat hair can be transformed into garments.

Transforming goat hair into lovely scarves and different garments has consistently been a training since Biblical occasions. Numerous individuals incline toward garments made of goat hair since they are light and are a superb warmth protector. They give this glow if vital during cold days, yet stay cool during the hot season.

4. Goatskin/skin has numerous employments.

Goatskin can be transformed into delightful embellishments, sacks, shoes, as a feature of furniture and enhancements of dividers and focal points.

There are different employments of goats other than those referenced previously. For instance, compost can be changed into moment manure by following natural planting rehearses. You can utilize your excrement for your nursery or sell it for benefit. Also, since these are herbivores that eat for the most part grass, they can be capably used to expel backwoods land by eating weeds.

How to Raise Goats

You'll be pondering where you'll discover all that you have to think about how to raise goats. You can discover books on all parts of goat reproducing in many book shops and libraries or even on the Internet. There are additionally web gatherings on this point. You will discover numerous individuals with involvement in the data they are happy to share. You can pose inquiries and offer guidance on rearing goats. You will get familiar with a wide range of tips to guard goats sound and.

To begin with the correct foot, here are a few hints to support you.

Goats need a couple of sections of land of grass for appropriate nibbling. You have to perceive what your goat eats since they are a little fastidious about what they like to eat. Use and keep the suggested goat grain by hand and heaps of new or protected green feed to enhance the grass varying. The virus necessitates that they get significantly more food than expected.

They'll require a great deal of safe house to keep them warm when it's virus.

It will likewise be important to give a great deal of water. Five liters of water isn't remarkable for a goat to drink.

Programmed working dishes, containers or drinking bowls for water goats and different animals. In the event that you can not control goats consistently, programmed watering gadgets are a great idea to have. Ensure it is done in a manner that can't be solidified.

Goats don't do well when they're wet. In terrible climate, your goat comes out of the virus. Sheets should be changed like clockwork, it is new and clean. This is particularly evident in stormy or cold seasons.

Goats are continually searching for something to get. They'll investigate all that they can get. Fencing is an unquestionable requirement and must be goat-confirmation. Goat wall ought to be shielded from enormous pigs and ponies. Try not to go modest when you make a goat fence. Protecting your goats is the primary objective of a decent fence.

Protecting

To raise goats, the main thing you have to do is fabricate a pen to hold your goats. A four foot high work fence will function admirably.

It is sufficiently high that a goat doesn't bounce on it or get into it. High quality electrified barrier is a decent decision for goats; it would appear that smooth wire hanging solidly. It is moderately cheap, simple to construct and has a long assistance life.

The size of your pen relies upon what number of goats you need to raise. It is better and fitting not to put multiple goats on a section of land of land.

This gives them a lot of room and land to touch, albeit a few varieties are generally bigger than others and accordingly require more space.

For a littler goat, similar to a Nigerian Dwarf, will make a house for canines. For the floor, the ground floors are fine, and numerous individuals lean toward them in light of the fact that their abundance pee will be ingested into the floor and will require less sheet material, along these lines, simpler consideration.

What to take care of them

Contingent upon the specific variety you have as a top priority to raise, you will just need to take care of them roughage and an enhancement consistently.

Water is by a wide margin the most significant supplement. Animals can invest a great deal of energy without food, yet not without water.

Goats typically drink ½ to 1½ liters of water every day. An awesome counsel on the most proficient method to raise goats is to take care of them for the most part roughage. Feed is produced using spices or vegetables that have been cut at a beginning phase of their turn of events and left to dry in the sun.

It must be green and be uniquely developed grass for food, for example, Garden grass, horse feed and Meadow fumble. The nature of roughage is controlled by how well it is thought about in the field. On the off chance that it downpours while the feed dries, it loses its dietary benefit and could shape itself.

Be that as it may, then again, if the feed dries excessively, the nourishment will be broken and lost during the pressing procedure. The best feed for the consideration of goats is horse feed or Clover streams, however don't take care of them white clover as it is harmful because of its elevated level of protein and calcium.

Another alternative is to take care of them with grains and they could eat the greater part of the rest of the plant nourishments. They will likewise keep their yard cut by eating grass.

Proliferation Management

Goats are generally occasional raisers, which implies they go into heat (estrus) and breed in the fall as the days abbreviate. The typical rearing season for most goats stretches out from August to March, albeit a few varieties and people breed consistently.

The Gestation of a deer keeps going around five months or 150 days. Typical conveyance happens in the nose between the two front legs. After birth, youngsters should take antibodies to keep them fit and solid.

For those of you who like little subtleties, when you have chosen to raise goats, you have to ensure that lone all around adapted goats are raised. They ought to be sound, very much took care of, however not fat since fat causes goats a great deal of difficulty imagining and kidding.

Numerous goat raisers work on "washing", which means checking the nourishing admission before the rearing season. This training keeps ISPs in ideal conditions for proliferation and causes them produce ideal eggs during ovulation, giving them more opportunities to imagine more.

Choice methods utilizing the most alluring goats or goats of decision for the specific reason for rearing. On account of the choice, the variety is improved. The rearing of goats should be possible by inbreeding or ancestry, by intersection and by intersection ancestries.

Inbreeding requires mating of related goats. This outcomes in a uniform posterity in execution and appearance.

The cross between two types of goats. Its focal points incorporate expanded regenerative productivity and power. Going too far methods crossing a few bloodlines in a single race.

The conclusive outcome of goat rearing eventually relies upon what you need and the technique picked to accomplish it. Furthermore, except for intersection, a definitive objective is to acquire a predictable posterity from age to age.

Economy

Produce goat meat can be beneficial, yet give unique consideration to subtleties. It is important to control flexibly costs and keep up significant levels of creation. Along these lines, the expense of raising your goats will be moderately low and yields will inevitably be high.

Presently you see that raising goats is a delight. I have arranged these elevated level advances and tips for you to find a way to raise goats without any preparation without the dread of committing expensive errors en route.

In this way, in half a month, you will no doubt have the option to get results. Goats require negligible venture and high return. At the point when you sell your goats you can hope to get from $ 50 to $300 for every goat.

RAISING GOATS - HOW REARING GOATS CAN HELP BRING IN ALTERNATIVE INCOME

With spending ever increasing, you need to be smart to find ways on how to generate revenue. A farm, the breeding of goats, has since become a popular way to be financially stable. Unlike poultry or other animals, goats are relatively easy to care for. They can survive with minimal monitoring as long as they are protected from the weather and receive enough food and water.

How to make a living raising goats? First of all, it is a fact that goats are an important competitor in the field of dairy production. Goat's milk is more favored by many than cow's milk because the former is easier to digest. So, goat's milk is more recommended to drink in children and sick people. You can also capitalize on milk by supplying it to milk processing plants. Soaps and lotions were also made with goat's milk, which only expands the market.

If you are not going to breed goats for milk, making money from their meat is an alternative. Goat meat is more expensive than pork or beef, especially if you have the best breeds. You can also sell the wool obtained by mowing goats. Such a fiber commands a high price on the market. If you decide to focus on the income of wool, needless to say, you need to make additional efforts to care for goats to make sure that they do not suffer from illness.

Some people, even tempted to start breeding goats, doubt the success of its execution. They think that people who have lived in the countryside have a better hand in running a business like goat farming. Contrary to this belief, everyone can be experienced in goat management, as long as they are ready to learn the entry and exit of the industry. Yes, this is not a joke, because you will have to shell out a considerable amount, but if you are endowed with training and knowledge, nothing will be wasted.

You can start by evaluating your first position. Is it suitable for goat breeding? If not, then maybe you can scout for a better home for your pets. This is not necessarily a vast open field, but a place where the goats will be comfortable. Better start with a few goats and a barn wide enough for them. Maximize your resources like food like hay or grass to reduce expenses. Make friends with people who have been raising goats for a long time because they can share best practices with you.

RAISING GOATS - LEARN THE FACTS BEFORE SETTING UP A GOAT FARM

With regards to making your own goat ranch, information on goat cultivating is certainly the key. On the off chance that you need to have the option to raise sound goats, you should initially set yourself up by gathering all the data you need.

Goats can be captivating animals to think about, and as long as you have all the fundamental apparatuses, their consideration and support can be simple. These are only a portion of the realities you have to think about how to raise goats.

Raising goats certainty # 1: unique types of goats require various degrees of care

In spite of the fact that it is anything but difficult to exchange with one another, you have to comprehend that there are in excess of 300 varieties known today, the greater part of which are reproduced to create milk, meat or hide for the present customers. Dairy breeds, for example, Alpine goats require an alternate kind of care than boere goats that produce meat regarding food and enhancements. Knowing the various types of goats, it will be a lot simpler for you to choose which explicit sort to decide for your goat ranch.

Raising goats actuality # 2: goats are extremely inquisitive animals

Something else you have to know is that they are normally extremely inquisitive about what encompasses them and furthermore have a significant level of knowledge. This is presumably why they are known to bite on all that they see around them. Regardless of whether it's jars or PCs, goats are glad to bite an assortment of things. Along these lines, before requesting goats for your homestead, ensure that nature is liberated from plants and articles that can hurt your wellbeing when ingested.

Raising goats actuality # 3: picking a subject matter

At long last, before you begin reproducing goats, you should initially choose what sort of item you need to create. Pick a subject matter and spotlight on it. Numerous amateur goat reproducers regularly tragically want to raise a wide range of goats simultaneously, in light of the fact that they accept that the consideration of goats continues as before. In any case, this circumstance once in a while works effectively, on the grounds that such goat reproducers rapidly understand that it is substantially more gainful to concentrate on a specific stock without a moment's delay.

Raising goats can immediately turn into a truly productive exertion, as long as you put a push to figure out how to appropriately think about your domesticated animals. Continuously get to know the most recent improvements in the protection of goats and you will doubtlessly prevail in a couple of months.

RAISING GOATS - HEALTH TIPS TO HELP YOU IN KEEPING GOATS SUCCESSFULLY

On the off chance that you are one of numerous individuals who are pondering raising goats, you should know by this point, raising goats isn't a simple undertaking, particularly in the event that it will be the first occasion when you will have the option to encounter it. In any case, regardless of whether this is the situation for a great many people, it ought not imply that you can not plan something for encourage the procedure. Another issue that people when choosing to raise goats on their homestead is the way that they really don't have an away from of how the choice procedure ought to be finished. Truth be told, just a couple of individuals who as of now breed goats are very much aware of what they do with regards to controlling the strength of goats. Hence, this article will give you some wellbeing tips that you can use to truly figure out how to raise goats.

As a matter of first importance, you ought to consistently recollect that the food you give your goats is extremely critical for knowing how they will come out when they are completely evolved. All things considered, a significant detail to remember is the way that offering food to goats that have too moderate isn't acceptable. Concerning people, moderates can get destructive to the soundness of their goats, since this may likewise be the reason for their passing. What's more, you ought to likewise remember that in the event that you purchased your goats, it is imperative to chat with the past proprietor, the particular eating regimen for which goats are utilized. Realize that occasionally changing the eating routine of a goat will prompt various sicknesses. Then again, on the off chance that you find that your goat simply needs to eat basic kinds of nourishments that typical goats eat like grains, feed and grub, you should fuse them with water to encourage absorption.

You should realize that goat rearing ought not be a troublesome procedure for you, regardless of whether you're only an apprentice; you simply need to recall that you should simply tell all of you of the fundamental data that you should have the option to prevail at long last.

THE MOST EFFECTIVE METHOD TO START REARING MEAT GOATS CORRECTLY IF YOU DO NOT HAVE ENOUGH EXPERIENCE

On the off chance that you are going to begin reproducing meat goats, however you just have questions since you need more involvement in it, this article will attempt to give you data that you can use all the while. Probably the best strategy you can use to figure out how to develop goats is to converse with somebody who should as of now have a great deal of involvement in them; for instance, it would be extremely useful in the event that you visited your ranch every now and then and by and by watched the procedures in question. Notwithstanding gaining from individuals, you can likewise get a great deal of valuable data from the different courses, exercises and shows that are accessible more often than not. At last, there are a few sites that you can likewise gain so much from most, particularly in the event that you are truly searching for significant data that you can utilize.

In the wake of finding however much essential data as could reasonably be expected about rearing goats for meat, the following thing you should consider are the diverse variety classifications that you might want to remember for your rundown of goats for reproducing. For instance, you really need to choose whether you need to increment enlisted rearing stocks or whether you need meat goats for business utilize that are not enrolled. You ought to likewise solicit the various sorts from breeds you need to raise; on the off chance that you are inexperienced with these; there are a few sources that you can investigate monitoring the different kinds of breeds that you can think about reproducing on your ranch.

Subsequent to knowing all the things you have to think about the accessible varieties you can raise available, there is something else you ought to do before you truly consider rearing goats for meat and this is to choose the quantity of goats you need to begin. To effectively decide the correct response to this issue, there are a few things you ought to consider, for example, where you will live or your sanctuaries, fences that will hold you together, and pastures that you can touch. On the off chance that you don't as of now have this hardware, it is critical to know first where you can get it before choosing what number of goats will start to develop. Other outside elements that you can consider to decide what number of goats you have to begin developing are things like the atmosphere, the specific kind and fruitfulness pace of the dirt, the slant of the dirt, and the sort of vegetation found in the dirt.

Having considered every one of these elements that have been referenced in this article, regardless of whether you're just a fledgling, you don't need to stress over raising goats, veal, since you will have a simple chance to have everything that you must have the option to have accomplishment with your business.

Caring For Goats

The consideration of goats is a genuine duty. Goats need organization. Consequently, it is smarter to have two goats or to have an amiable creature to give just a single goat organization. Goats need extensive havens, liberated from dampness and drafts. There must be acceptable ventilation with the goal that the newness of the air is kept up inside the sanctuary.

Horse shelters with three sides and rooftop are perfect for every climate condition. Wall for haven or compound safe house ought to be higher than goats will in general ascension. For bedding, dry straw or dry wood chips. Ensure that wild animals or canines can't enter the safe house, as these animals can hurt goats.

It requires uncommon thoughtfulness regarding taking care of goats. Goats don't lean toward messy food. It is ideal to take care of goats with grub, for example, route and feed, grain nourishments and dietary enhancements. They are touchy to abrupt changes in diet. Make changes to the taking care of calendar as for taking care of time, taking care of type, and taking care of sum, step by step.

The water offered to the goats ought to be cleaned and conveyed in clean holders. Overall, goats can devour two to five liters of water each day and this relies upon the variety and size of the goats. In warm climate, it is acceptable to offer water at shorter spans, and when it is cool, high temp water is suggested.

Access to a veterinarian routinely will help identify any illness or disease. In the event that goats show changes in their dietary patterns or other everyday practice, it is ideal to do as such. Consistently cut off goats ' hooves and deal with them if there should be an occurrence of disease. Spot the flycatchers inside the sanctuary, in light of the fact that in the warm season, flies will in general pester goats. Shave the goat throughout the mid year if the hotel is hotter. Worm goats use vermifuge glue in any event once per year.

The season reaches out from the last piece of summer until the start of winter is useful for goat reproducing. There are eighteen to twenty-one patterns of estrus for The does'. Goat raisers utilize characteristic proliferation or managed impregnation. - I become ripe at an early age of two months. Solid goats and goats seven months old enough or more established can bring up and give sound kids, principally twins. A deer takes five months to conceive an offspring.

Goat cheddar is turning out to be mainstream nowadays as goat cheddar has great dietary benefit. For the creation of goat cheddar, goat milk, whey, scoop, new lemon juice, sifter, cloth, and so on are required. To make goat cheddar, blend all the fixings in the bowl.

Subsequent to emptying it into the dish, heat it up to 170 degrees, at that point cool it for twelve hours normally in the wake of covering it with a saran wrap. Channel the cheddar blend through a sifter in the wake of putting the cheesecloth. When depleted, expel the dressing and store in a sealed shut compartment. Making goat cheddar is conceivable at home.

When all is said in done, goats are significant animals. Regardless of whether it's your milk or your skin or your fat or your meat or your dung, everything has esteem.

Rundown OF GOAT FARMING BASICS

Prior to managing your first pair of goats, ensure you have the correct hardware and offices to think about them. Beneath you will discover the essentials of goat cultivating. Think about them, check on the off chance that you have all the necessities and begin raising goats as pets or for business.

Obviously, the primary thing you need is sufficient field. We suggest a couple of sections of land of grass to eat a goat the more goats you intend to get, the more space you need. He needs it to have the option to take care of his goats appropriately and give them sufficient space to move around.

Ranchers for the most part feed their goat feed, yet they can do similarly too with horse feed supplements. There are additionally a few brands of goat grain accessible. Simply pick the sort and brand that is generally appropriate for goat reproducing and you are all set.

You ought to likewise guarantee that your goats approach drinking water. You can utilize a container in the event that you don't have a watering can. Make certain to fill and clean the can each day. You can likewise have a programmed water station introduced in your animal dwelling place. Along these lines, ensure your goats have something to drink and that the water they get is perfect and new.

Choose what kind of lodging you can make for your goats dependent on your financial plan and inclinations. An outbuilding will be more valuable as it will give your goats the assurance they need. You additionally need to fabricate tough wall around your grass to shield your goats from predators. Ensure it's sufficiently high to keep the goats inside. Ranchers suggest that wall be in any event four feet high, made of twisted wire and upheld by posts divided in any event 12 inches separated.

With these prerequisites set up, raising goats (both for business and for pets) will surely be a pleasant encounter. Simply ensure that a veterinarian normally checks the goats to keep them from getting a malady.

TIPS TO KEEP GOATS HEALTHY ALL YEAR ROUND

Goats are acceptable pets and livestock, however you have to realize how to appropriately raise a goat on the off chance that you intend to keep them solid consistently. Get familiar with the regular predators of goats on the off chance that you intend to raise on the Meadows. Then again, in the event that you need to take the way of pen rearing, you are as of now honored with a few advantages to do as such. Having goats in a confine is better since you can undoubtedly watch. Taking care of them is likewise simpler in light of the fact that you can place them in one place and never need to stress over getting them. It additionally gives better haven from components and vermin. A decent sanctuary will get your goats far from pneumonia, which is a typical malady among them.

When planning the goat pen, ensure it is sufficiently simple to clean. Lift the floor a couple of centimeters from the floor to make upkeep basic. At the point when you need to raise later, it is smarter to have a bigger space for them to move around. You can likewise fabricate a different pen when your goats are breastfeeding and furthermore to maintain a strategic distance from goat scent moves. The most effective method to raise a goat is simple in the event that you follow these first arrangements. With regards to Goat Milking, make certain to have a cognizant and pre-arranged program to follow carefully. Twice-every day draining is adequate and ought to maintain a strategic distance from delays in the arranged draining periods.

Continuously wash hands and other draining devices while draining goats. What's more, it is fundamental to take care of your goats something during draining so they are loose and feel compensated during draining movement. With respect to kids growing up, they are constantly housed in a different pen. Check the overabundance areolas in young ladies and have them evacuated as quickly as time permits in the event that you need to raise your milk. For meat creation, guys must be mutilated inside the main month; this is the correct method to raise a goat.

Moving currently to do as such for propagation, they as a rule arrive at pubescence somewhere in the range of four and year and a half. They can deliver 2 kids in a single brood. Limit generation to once every year to keep youngsters sound. Then again, youthful dollars ought to be constrained to serving 25 every year. The greater your cash, the more it can serve. Follow this guide on the most proficient method to raise a goat and you are rarely off-base.

Goat Meat Nutrition

At the point when your goat cultivating business has quite recently begun, you will require all the assist you with canning get for your fruitful and beneficial business. In the rearing of goats, you have the chance to go into the creation of milk, meat creation, even from the offer of their posterity. In the event that you choose to wander into meat creation, there are a few contemplations to consider, most importantly, the sustenance of goat meat.

When raising goats for meat creation, it is important to guarantee that the crowds are thought about, as far as goat lodging, yet it is likewise imperative to deal with the wellbeing and government assistance of the animals.

Given the sustenance of goat meat, it is critical to remain centered when taking care of these animals. Goats may turn out to be excessively touchy to the eating routine you give them. It is your duty to guarantee that crowds get the best in taking care of on the grounds that outcomes, for example, demise can be set off by lacking taking care of.

The nature of the sustenance of goat meat will to a great extent decide the development of crowds, the quality and amount of milk creation, the growth and strength of its sources. Any reasonable person would agree that goat reproducing relies just upon how these animals feed.

There are a few variables to consider while picking goat feed. It would be more helpful on the off chance that you allude the inquiry to somebody who has understanding and has the ability to discuss wellbeing and sustenance, veterinary.

Goats, similar to people, as different animals, must have a reasonable and wholesome eating routine. Your eating routine ought to contain a lot of water, nutrients and minerals, vitality and water. Make certain to counsel a veterinarian before taking care of your crowds.

We can give you 3 Simple Tips to get amazing sustenance from goat meat.

1. Nourishments that contain unnecessary additives can surely be unsafe to groups. You'll never know how these animals will respond to food additives.

2. In the event that you have as of late gained a goat or groups, you should converse with the previous proprietor of the eating regimen to which the goat was oppressed. This will guarantee that there will be progression and

hypersensitive responses, even demise brought about by abrupt changes in your eating routine will be likely.

3. Despite the fact that nourishment of goat meat is a significant factor in goat cultivating, it is consistently imperative to remain on the spending plan, thusly you are guaranteed of acceptable benefits. There are productive nourishments you can think about, for example, roughage, wheat and feed. They should likewise have a sufficient flexibly of drinking water.

Sustenance of goat meat is a fundamental piece of goat rearing, so additional safety measures ought to be taken during taking care of.

A Quick Guide to Goat Health

Goats get a considerable lot of similar infections that happen in any dairy creature, so much consideration ought to be paid to the strength of goats. The most ideal approach to keep your goat malady free is to keep everything spotless and sterile. Guarantee that a veterinarian every now and again checks the wellbeing of the crowd and stays up with the latest.

Something you have to do is fundamental for the strength of goats is to build up an arrangement for bother control. Your goats regularly purify and secure against outside vermin, for example, ticks, lice, mosquitoes and a wide range of flies: House, Horn, outbuilding, pony and deer. Every one of these gnawing and sucking creepy crawlies can influence the state of their goats, in some cases causing looseness of the bowels and influencing the creation of milk. You should converse with your neighborhood district farming official to perceive what are the most ideal approaches to take out these bugs.

Mastitis is an ailment that influences dairy goats and dairy bovines. This truly implies there is irritation of the chest. While analyzing the udder may seem tense, hard, hot and cause extraordinary uneasiness to the goat. Both intense and ceaseless mastitis is treated with anti-infection agents.

The principal line of safeguard against mastitis is cleaning during draining. Creatures that cause mastitis might be available in the earth. Fertilizer more be expelled every now and again. All draining gear, both with hands and with machines, should consistently be clean. An answer dependent on chlorox fade (MC) ought to be utilized to clean the areolas.

Bosom edema is another medical issue of the goat. This normally happens in dairy goats towards the finish of their dry period. It is treated by controlling the measure of sodium, potassium and corn flour in the goat's food.

Abscesses (caseous lymphadenitis) can be an issue in grown-up goats and can even prompt demise if the abscesses encompass an inner organ. They are regularly found on the neck, shoulders or head. The condition can be treated by depleting the abscesses, cleaning the territory and controlling goat penicillin. In the event that this issue creates, the influenced goat or goats ought to be disengaged to forestall their spread to the remainder of the group.

On the off chance that goats live in a territory regularly moist, rottenness of the foot is a chance. There will be a dark, rancid release, and the goat will be weak and agonizing. It is treated by cutting the decay and applying an answer of copper sulfate or balm. A decent cut of the caps can help keep away from this issue.

The best arrangement for goat wellbeing is to screen your goats consistently with the goal that you can rapidly see any adjustments in stance or conduct that could be a sign that something isn't right and a veterinarian should be called. The better you know your animals, the simpler it will be to promptly observe that something isn't right or wrong.

Individuals are associated with the consideration of goats for a wide range of reasons. Regularly goats are reproduced to give milk or meat, mohair or cashmere, to clean the land or become pets. There are around 7,000,000 goats on the planet, the greater part of them in creating nations where they are utilized for milk creation.

Goats are extremely solid animals, so with regards to thinking about goats, they needn't bother with a great deal of lodging. For the winter there ought to be a sanctuary with dry sheet material. Goats despise getting wet, so cover from downpour and snow is significant. A three-sided cover functions admirably lasting through the year as it shields your goats from the downpour. They generally have in any event two goats together in light of the fact that goats are soma animals and can be forlorn and miserable on the off chance that they are the main creature.

Goats should be fenced and still, after all that they will attempt to escape now and again. At that point you have to get a decent tough fence with a durable wire. An electrified barrier can work very well until you have canines or different animals sufficiently little to sneak under the fence and get into the fence or grass.

With respect to food, thinking about goats implies taking care of value roughage and wheat consistently. They should likewise get minerals and have new, clean water consistently. Ensure that the feed doesn't stay on the ground where it can get messy and wet. A wooden roughage backing ought to be worked for taking care of the dirt. Along the fence line, it very well may be the perfect spot since when you put it in too little a space, the predominant goats can forestall the less prevailing or more youthful ones from arriving at the feeder.

Goats ought to be brought steadily into new fields or newly cut green roughage with the goal that they don't grow and don't become ill. While you can hear that goats eat everything, actually, this is a long way from valid. Goats are, truth be told, exceptionally requesting with the roughage they will eat. A few proprietors of goats feed horse feed and Meadow sole with great outcomes.

Goats should be immunized consistently, so on the off chance that you start with goats, you should locate a decent veterinarian of huge animals. They additionally should be dewormed consistently and have stops up cut off. Despite the fact that goats are regularly not wiped out, it regards search for a little goat maladies with the goal that you recognize what manifestations to search for and when to call a veterinarian.

On the off chance that you choose to raise your goat, the female and the male must be in season simultaneously. It is as a rule in the fall, among August and December. During this period, a female goat heats up each 18-21 days. Subsequent to being raised, the incubation time frame is around 150 days, The Young will be conceived in the spring.

TAKE YOUR GOAT'S TEMPERATURE

Some of the time your goat may not act ordinarily, in such cases the principal thing you have to do is take its temperature, since this is the primary thing your veterinarian can ask when calling the treatment. Regardless of whether the goat doesn't work from a high or low temperature, it would give an insight to your veterinarian about what the genuine issue may be. The typical temperature of a boera goat is 101 ° F to 103 ° F in winter, spring and even harvest time. It is standard for your goat to lounge in the sweltering sun the entire day and have a temperature of 104°F, anyway it must fall rapidly once the goat comes out of the daylight.

Taking the temperature of a goat is equivalent to taking the internal heat level of a human youngster. You can utilize advanced and conventional glass thermometers, which can go from $ 3 to $ 6. Glass thermometers have a solace ring at the external end to tie a rope. You should be cautious whenever utilizing it as it has more possibility of breakage. For Boer goats, the most agreeable position is to put them on their knees. Try not to drive the thermometer into a goat, it should slip effectively when greased up. Greasing up the finish of the test with oil or petroleum jam would assist with encouraging addition into the rectum. The situation of the thermometer ought to be set on a large portion of its length and hold it set up for in any event two minutes. Advanced thermometers would give a caution when it is through.

For more seasoned youngsters, it's acceptable to have another person holding the goat in a vertical or slanted situation, since you can take its temperature. For grown-up boere goats, particularly those that are not restrained, they might be simpler to connect with a link or neckline and an info post has rope, they may likewise require somebody's assistance to balance out them until you embed he svietidlo and get its perusing. On the off chance that the temperature of the Boera goat is high, the veterinarian may recommend Banamine by infusion to decrease fever and agony. For Boer goats, it is smarter to squash ibuprofen into powder and blend it in with a little water. This assists with lessening the stifling element of a dry pill. It is smarter to beat with a high temperature, yet it is fundamental to discover the purpose behind the fever. A recognizable reason for exceptionally high temperatures is respiratory contaminations (now and then pneumonia). In the event that you need to spare your goat, you have to treat both the temperature and the contamination. A few anti-infection agents are truly available through goat flexibly stores-the most much of the time utilized items are oxytetracillin and penicillin. A few reasons for contamination ought to be treated with more costly solution anti-microbials, for example, Nuflor or Naxel. It is smarter to approach your veterinarian for the correct item and measurements sum, as opposed to just trying different things with over-the-counter items. We can not finish this article without referencing two additional things about the temperature of the goat. The first is that high fevers lead to lack of hydration. It is essential to oversee electrolytes to keep the goat hydrated. Furthermore, a Boer goat infant with high fever ought to be expelled from the milk until the fever is relieved, until the GOAT is hydrated with electrolytes. Electrolytes are significant for the capacity of organs and muscles, blood stream and the evacuation of fluid waste.

Precaution

A thermometer should be cleaned properly with an alcohol swab after each use and fixed in its case. Do not use a dirty thermometer - although several goats seem to suffer from the same kind of problem. Do not make the mistake of inserting an impure rectal thermometer into the vagina of a deer. Thermometers should be stored at room temperature. Glass thermometers should be "shaken" before and after each use. Digital thermometers must be rearranged according to the manufacturer's instructions.

Goat Diseases

Talking about a wide range of sicknesses of the GOAT, I found that numerous individuals don't know about the physiological databases that must go with the right conclusion.

All things considered, here are a few realities that we will examine in this area about goat infections. Fantasies about people contracting goat infections. Assessment of goat infection, assessment agenda and complete rundown of all goat illnesses. Covering the most well-known and furthermore, the rundown of uncommon ailments of the goat. I am gathering a total rundown of goat sicknesses that will be distributed on the goat fellow site in a downloadable record group.

Typical Physiological Information Of Goat.

Temperature: 104 ° F

Pulse: 70-80 every moment (kids are quicker)

Breathing: 12-15 every moment

These are the essentials you have to know to keep your animals solid. Maladies of goats are appropriately overseen by crop turn, dispose of congestion and sufficient medicine.

Assessment of Symptoms of Goat Disease;

You ought to watch every one of your animals somewhere around consistently for a visual correlation with decide whether there are any medical issues. You are searching for manifestations, postponed crowd, absence of appetite, limping, loose bowels any surprising conduct. You are additionally searching for teeth that squeak and snarl. In the event that you feel that you need a Professional Examination, contact your veterinarian. The person in question has to comprehend what your undeniable social contrasts are between the talked with crowd and the ordinary group.

Update of the Checklist;

How about we think about the age of the creature viable. You can remain in his place. He has his own vision or falls on objects. Look how it harms. The goat looks swollen and snorts, or the swollen region checks breath every moment (here you are searching for growing and exhalations of the tummy appears to have the runs it is normally evident that it has swollen udder which is everything you can outwardly see from a short separation. It will at that point be important to look at the others. You have to connect with your pet. Be cautious here, you don't have to practice or perform it will influence the temperature, breathing and heartbeat perusing. That we'll need to make an appraisal.

Take the temperature of the goats. Thermometer in the butt-centric pit. Tallies the Heartbeat (beneath the lower rib) thumps every moment controls the eyes for vision issues, items, running or in the event that it streaks with the development of the hand, gradually towards the eye. Test around with your palm and feel for thundering development. Note: If you feel agony or sloppy or loaded with water around there. Tune in to the chest region for clatters, wheezing. Stethoscopes are effectively accessible. Put your head against your chest and tune in. Check the mucous organs for pink or practically white shading. Doe lactation: check the chest for expanding, protuberances or hardness in the chest. Check the consistency of blood and milk in milk. Feel the warmth on the udder. It is somewhat an exhaustive assessment, and with the data gathered, it is conceivable to make an educated evaluation. The veterinarian should realize these things to help you more without taking the goat.

Shielding YOUR GOATS FROM COMMON ILLNESSES

Like other livestock, goats likewise experience the ill effects of different sicknesses, particularly in the event that they don't take incidental inoculations. These goat illnesses can be bacterial, viral or parasitic, among others. Since a large portion of these ailments can barely be resolved in view of their indistinguishable side effects, it would be attractive for the proprietors to look at them occasionally to serve the goat and the wellbeing of its misuse.

Ketosis (ketonemia) is one of the most widely recognized illnesses of the goat because of the creature tendency of concentrated food. Another difficult ranchers face is the point at which the GOAT turns out to be profoundly plagued with worms. See it can tell when the GOAT has a huge head on the off chance that it abruptly builds up a swollen head joined by whiteness of the skin. This ought not be mistaken for expanding, another regular issue with goats. At the point when you take a gander at your goat, you will see that it is frequently peeing, on edge looks and acts uniquely in contrast to the rest. Right now, the goat ought to be taken for conference to keep away from intricacies.

Sicknesses in goats can be forestalled by ordinary checks. A visit once every month to the veterinarian would be attractive. Being mindful to potential issues inside the group would help forestall the passing of a goat. There are approaches to know whether something isn't right with one of the goats.

- Note which crowd is regularly disengaged. Goats are normally amiable and neighborly, so they cross a pet today and the division of the group would mean an indication of difficulty

- Look for variations from the norm in the stool and stool. As a rule, the individuals who experience the ill effects of an infection would have abnormalities in the stool or pee. Among the most unmistakable side effects for goat ailments are the runs (with blood), discharge from the ears, mouth, vulva or any piece of the body of the goat

- Swelling of the jawline is additionally another sign that your goat is debilitated

- Feel deserts in developments notwithstanding conduct

In the interim, you can likewise have control of your goat to ensure the group is in pink. The principle regions are rectal temperature of in any event 39 degrees and pulse which ought to be 80 beats or less every moment.

It is significant for each rancher to know the diverse goat ailments that can influence an individual from the crowd. Knowing the side effects and emergency treatment medicines would likewise help limit the hazard. Give them legitimate consideration of the goat, and you'll simply need to stress less.

Kind reader,

Thank you very much. I hope you enjoyed the book.

Can I ask you a big favor?

I would be grateful if you would please take a few minutes to leave me a gold star on Amazon.

Thank you again for your support.

Andrew McDeere

www.ingramcontent.com/pod-product-compliance
Lightning Source LLC
Chambersburg PA
CBHW080452220526
45465CB00006B/2247